# The Far-Off Land

# The Far-Off Land

## An attempt at a philosophical evaluation of the hallucinogenic drug-experience.

by

Eugene Seaich

1959

*Ich weiss nicht, was soil es bedeuten,*
*Dass ich so traurig bin;*
*Ein Maerchen aus alten Zeiten,*
*Das koramt nir nicht aus dem Sinn . . . .*

*(I know not the meaning of this melancholy,*
*A legend from long ago*
*Keeps running through my mind . . . .)*
*(German Folksong)*

Library of Congress Control Number:        2012912674
ISBN:            Hardcover            978-1-4771-4397-1
                 Softcover            978-1-4771-4396-4
                 Ebook                978-1-4771-4398-8

**To order additional copies of this book, contact:**
Xlibris Corporation
1-888-795-4274
www.Xlibris.com
Orders@Xlibris.com
118511

. . . need I dread from thee
harsh judgments, if the song be loth to
quit
Those recollected hours that have the charm
of visionary things, those lovely forms
and sweet sensations that throw back our life,
and almost make remotest infancy

A visible scene, on which the sun is shinin
(Wordsworth, *The Prelude*)

*Und nach so leidgetrSnkten Jahren*
*Die so vieles uns zerstttrt,*
*V/ird der Welt, die wir einst waren,*
*Sage imraer noch pehbrt.*
*Ihre Runen werden bleicher,*
*Ihre Tone fern und zart,*
*Doch sie hat in zauberreicher*
*Anmut ewige Gepenwart.*
And after years of suffering
Which destroyed so much in us,
There will always sound the legend
Of a world which once we knew.
Its runes are pale and faded,
Its music faint and far;
Yet its presence lives forever
With eternal magic charm.
("Hermann Hesse," Spate Gedichte)

*J'ai longtemps habite sous de vastes portiques*
*Que les soleils marins teignaient de mille feux,*
*Et que leurs grands piliers, droits et majesteuex,*
*Rendaient pareils, le soir, aux grottes basaltiques.*
*Les houles, en roulant les images des cleux,*
*Melaient d'une fapon solennelle et mystique*
*Aux couleurs du couchant reflete par mes yeux.*
*C'est la que j'ai vecu dans les voluptes calmes,*
*Au milieu de l'azur, des vagues, des splendeurs*
*Et des esclaves nus, tout impregnes d'odeurs,*
*Qui me rafralchissaient le front avec des palmes,*
*Et dont l'unique soin etait d'approfondir*
*Le secret douloureux qui me faisait languir.*
(Once on a time I lived in might vaults
which ocean suns stained with a thousand gleams;
their straight majestic columns made them seem
as evening deep grottoes of basalt.
The billows, tossing images of skies,
mingled in a solemn mystic mode
their music's powerful harmonies which glowed
with their sunset hues reflected to my eyes.
And there I dwelt among voluptuous calms,
in the midst of azure, splendor, and the waves,
and the heavy perfumes of the naked slaves
who cooled my forehead with slow fronds of palm,
and whose only duty was to seek
the hidden sorrows that had made me sick.)
("Baudelaire," La Vie Anterieure, A Former Life)

# Introduction

When the Spaniards entered Mexico in the early sixteenth century, they found the natives using a family of strange new drugs, unlike any they had known before. The Aztecs had given one of them the name, *Teonanacatl* ("The Flesh of the God") in honor of its miraculous properties, which enabled one to see visions, to foretell the future, and to obtain supernatural revelation. Farther north, the Conquistadors discovered a cult surrounding a mysterious, turnip-shaped cactus, almost indistinguishable from the stones of the desert, which enabled its users to behold the secrets of the universe. Duly noting the properties of these drugs, the Spaniards succeeded in stamping out their use. Through the centuries, they remained all but forgotten, until in the last century settlers in the American Southwest observed that the cult of the peyote cactus still survived among certain Indians, although in an altered form. As they became converted to Christianity, these tribes adapted their worship of the magic plant to its celebration as a Sacrament; in its new form, the cult spread as far north as Canada and is today incorporated as the Native American Church of the United States.

Men of letters learned of peyote after the middle of the nineteenth century. Its properties were exploded by cultured Europeans, such as Havelock Ellis and Alexandre Rouhier, who discovered for themselves the marvelous visions that the drug produced. The German pharmacologist A. Heffter finally isolated an active chemical from the plant, which was called *mescaline*.

In the meantime, confusion reigned among historians regarding the descriptions of the other Aztec drugs. In 1923, Dr. Blas Pablo Reko discovered that the sacred mushroom, *Teonanacatl*, was still employed by Indians deep in the mountains of Oaxaca. Subsequent investigators established the identity of the *ololiuqui*, or the seeds of a narcotic bindweed, employed by the Mazatec tribe.

In 1943, a Swiss chemist, Albert Hoffman, accidentally absorbed through his fingers a minute quantity of a powerful synthetic substance, which caused him to behold the same sort of visions produced by the ancient American drugs. It received the designation *LSD-25* (for d-Lysergic acid diethylamide, the twenty-fifth of a series of chemicals being investigated). Since that time, a number of other synthetic drugs, capable of producing hallucinations, have been created, including *DMT* (diethyl tryptamine), *T-9* (diethyl tryptamine), and *adrenolutin*, which closely resembles the metabolic hormones of the human body. In 1959, the sacred mushroom yielded its secret in the form of the drug *psilocybin*, a relative of the tryptamines mentioned above, and its psychogenic ally active precursor *psilocin*. Most recently (1960), the *ololiuqui* has been found to contain an isomeric form of lysergic acid, *isolysergic acid amide*, which possesses the same properties as LSD.

Ever since mescaline became available, the close resemblance between its effects and the symptoms of schizophrenia has been noted. In a classic study of the "mescaline psychosis," Tayleur Stockings (1940) observed that both "paranoia" and "catatonia" can be produced by administration of the drug. Psychiatrists and pharmacological researchers have accordingly suggested that mescaline, LSD, and related chemicals might provide a clue to the nature of insanity. Many studies have been made, and continue to be made, in hopes that an ultimate cure for mental illness might be found.

But, other studies of the human mind await the application of hallucinogenic tools, studies that might prove even stranger and more illuminating than those of the pharmacological laboratory.

Deep within each of us, the past slumbers on. All of the patterns of our understanding lie buried in the unconscious memory, shaping our desires, our inspirations, and our dreams. It is these ancient memories, particularly those at the deepest level of the organism, that perpetually appear as haunting suggestions of a prior existence, or a higher reality, which prefigures our picture of human life. This vast residue of mental experience is what the Greeks recognized as the *daimon*, or the sense of destiny that drives our conscious energies toward their necessary fulfillment. As an active repository of intuitive knowledge, it integrates and guides our understanding of reality; whatever we know, or feel, or hope to attain is rooted in its primal soil.

It has seemed to me that the well-established properties of the hallucinogenic drugs might be well employed to enable us to explore this far-off land, which is in effect our subconscious mind. Were we to learn its secrets, we would better understand our own desires and the motives that drive us through life. Still better, the secrets of human history would perhaps be discovered as the eternal patterns of imagination that have shaped our spiritual existence. But, perhaps most important of all, to penetrate the well of the past might restore to us that visionary perception that we think we once possessed. Legend and myth are curiously persistent in their suggestion that the human race formerly enjoyed the delights of paradise; actually, I believe that this paradise has been fashioned perennially by each of us from his own recollection of life's initial innocence, and therefore awaits recreation from the depths of primal memory. If this is true, the strange drugs that the Indians left to us might prove to be the very Hermetic Secret sought after by the alchemists.

In the study that follows, I have attempted solely to analyze my own experiences with two of these drugs, LSD and mescaline. I have not avoided treating them subjectively, since this aspect of the experience especially reveals what is operative beneath the surface of the mind when hallucinogenically stimulated. A cardinal principle has guided my observations: The human mind stands behind all phenomena, organizing, integrating, and interpreting; the nature of its "abreaction" to experience reveals its inner functions, just as our tastes and prejudices reveal our personalities. This principle is not proposed in an extreme Berkelean sense as a denial of objective existence, but as recognition of the essential role played by our total past in experiencing "reality," according to the image we bear within us. Nor does the private nature of my experiments preclude a general application, since each of us is an expression of our race and culture; any study of literature or philosophy will show that the same motifs appear continuously in history, illustrating basic insights that we inherit from life: insights both universal and timeless because of the existential problems faced by all. Quite obviously, the hallucinogenic experience is not stereotyped by a single type of personality; the details that follow are only suggestive of certain imaginative processes involved, rather than their necessary psychogenic form. Thus, one might comprehend in them a picture of human consciousness in general; for the deeper one penetrates the subconscious mind, the more impersonal it becomes and the closer one approaches the state that existed before conceptual egotism drove us into our separate worlds. There are, indeed, sufficient similarities between the experiences investigated here and those recorded in both psychological journals and the world's great literature to suggest an essential agreement between all subconscious memories. Accordingly, the present study attempts to discover the broader realities that lie behind psychogenic phenomena and to seek a pattern that would explain the longing of human beings for the Beyond, for the otherworldly substance of their intuition. Whether or not we are successful, it is hoped that fruitful directions for further investigation will be perceived, and the use of our new hallucinogenic tools will be extended to much broader fields than is presently the case.

# -1-

We are told that, in those moments immediately preceding death, the world of our earliest infancy frequently opens up to us. We are also assured that senile reason passes readily into a state of second childhood, wherein the light of rationality is obscured by the resurrected past, experienced as fully as if the intervening years had rolled away. Normal adults occasionally dream of long-forgotten events, which have otherwise passed into oblivion. These facts, together with Freud's rediscovery of the unconscious mind, suggest that within each of us the past slumbers on, occasionally reasserting itself in the fragments of a sudden recollection, the perception of some haunting perfume, or the unexplained appearance of an ancient face in our dreams. Most striking, however, is the fact that it is this earliest layer of the human memory that persists to the moment of death, even after the adult memory and its powers of reason are gone. Knowledge recently gained disappears, while that mysterious world of the long forgotten is reawakened, showing that our earliest experiences indeed have vitality that rational intelligence does not possess.

The strange discovery in recent years of certain drugs that can open up this buried world at will seems to me to be worthy of our best romances, wherein men have ever dreamed of piercing the veil of memory in search of the ultimate secrets of Being. I can scarcely describe the excitement that possessed me when I first held in my hand a tiny vial of whitish powder, extracted from the sacred cactus, which the ancestors of the Aztecs worshipped millennia ago, when men believed in the Mysteries of Existence, now laid open to the dead knife of scientific analysis. Before me, in a heap of delicate needles, lay the divine power, which the ancient Indians identified with life itself, a supernatural and invisible force pervading the visible world, referred to by anthropologists as *mana*. Modern Indians says that God has sent His Holy Spirit in the form of the peyote plant, and he who eats thereof may take into himself this Power; he will see visions, obtain hidden knowledge, and be led to the Truth that evades his grasp. Even stranger claims have been made for the sacred mushroom, *Teonanacatl*, which can extend man's vision to the future, to the past, or to remote occurrences of the present. Today, a third such chemical has emerged from the synthetic laboratory derived artificially from the ergot fungus. All three substances have one thing in common: the power to penetrate that deepest layer of the human mind, that mysterious realm that lies beyond the veil of ordinary perception. The green tea with which Le Fanu's Rev. Jennings obtained glimpses of the celestial arcana, the tincture that dissolved for Stevenson's Dr. Jekyll the "primitive duality of man," or the prophetic vapors that issued from the earth beneath the oracular tripod at Delphi could scarcely have been more marvelous than these drugs we now possess, which are in fact mysterious keys to the inner soul of Man.

The existence of man's subconscious has been known for countless ages. Oriental psychology recognized it long before Dr. Freud redirected our attention to it. It has also been pointed out by the school of Jung that this subconscious operates through recurrent patterns of memory that are imposed upon our waking knowledge. Plato's "ideas," Swedenborg's "correspondences" (derived in turn from the medieval *correspondentia*), or Kant's "categories of reason" are but older varieties of this idea that human life is a reflection of some prior kind of knowledge, a knowledge that antedates the world of rational experience. Poets have longingly searched for this otherworld; whole races have created myths symbolizing their shadowy recollections of it. Men throughout history have employed in their sacred mysteries such drugs and intoxicants as they possessed, with the hope that they might regain a fleeting glimpse of the beyond. So universal is the human longing for the "otherworldly" that it constitutes an archetypal experience in itself. Whether or not it reflects the memory of the race, as Jung has suggested, or is merely the result of our cultural experience is immaterial. The fact alone explains how, in passing from life, the dying man finds the well of his childhood opened up once more . . . how, at death, he attains the Unknown by way of his earliest existence, an existence that is shown to be present during every moment of life; for if this were not the case, its persistence unto the very end would be impossible.

# -2-

Childhood is our first experience with life; it will therefore be obvious that our deepest attitudes are the product of a childlike intellect. This earliest vision is the one against which we must inevitably compare our present situation. Far back into the distance, a furtive memory stretches, like a far-off land, visible chiefly through the haunting suggestions of dreams. This labyrinth of half-forgotten existence is opened to us by a variety of methods, mostly gratuitous. In those strange moments when unexpected sights and sounds suggest past memories; when the bosom of nature seems to recall some childhood knowledge of flowers, trees, and clouds; or when a dream reawakens some ancient experience from out of the past, we feel as if life indeed posessed another dimension, a deeper level of prior existence, perhaps more real than the one that we ordinarily inhabit. Most importantly, the powerful charm of such a suggestion, complete with its memories of freedom from the burdens of adult necessity, invests this "otherworld" with the qualities of a primal paradise. The formative years of our lives are the ones most devoid of responsibility, watched over by parental authority suspended, as it were, in time and space, scarcely aware of the laws of causality. Something always rebels in the sensitive individual against the ugly demands of material life because the unconscious memory suggests a previous stage of effortless happiness. I am convinced that this experience lies at the root of the universal myth of lost Paradise, wherein perfection is equated with innocence, and good and evil are deemed to be penalties of adult knowledge. Such legends are found amongst all peoples: The Gaels dream of the lost Isle in the West, *Tir-na-n-Og*, the land of Eternal Youth, called Avalon by the Brythonic Celts. According to the Taost symbolism of the Golden Flower, the beginning of things to which the soul longs to return is a state of perfect oneness, unconscious of the opposition of good and evil, located in a sea of primal life-force, comparable to the Waters of Life that flowed forth from Eden. I would like to point out what may well be the real meaning of the Sumerian, Babylonian, and Hebrew myths of Paradise: Here again we find in poetic form a representation of our universal memory of lost perfection, fashioned from dim recollections of innocence, now obscured by analytical adult knowledge, i.e., the "knowledge of good and evil." What we once enjoyed as a simple experience, we are now compelled to analyze in light of complex social, philosophical, ethical, commercial, and scientific values, thus destroying the primal enjoyment of childhood's mysterious world. As Byron's Manfred tragically observed, "The tree of knowledge is not that of life." Consequently, history has witnessed the ubiquitous appearance of mysticism, which seeks to repair these shattered opposites of conceptual thought with the whole cloth of direct experience. Mahayana Buddhism, for example, describes this need for reintegration as follows: The superficial concepts of the causally oriented ego must be extinguished (Nirvana), so that pristine existence can again be experienced. Ultimate reality is what has always existed, be it only perceived without the compulsive anxieties of the worldly mind (Sangsara):

> Only let your mind dwell
> In the realm of Nothingness,
> And you shall see not far off
> Hakoya's Transcendent Hills.
> (*Manyoshu*, xvi, 3851)

Again, Christ likened the kingdom of Heaven unto the mind of childhood; this was not because children are easily beguiled, as orthodox religion would have it, but because Christ understood the world with which children are conversant. Children alone are capable of pure innocence, that they might consider the lilies of the field, which to the poetic mind are clothed in such beauty that "even Solomon in all his Glory was not arrayed like one of these." Thus, in the dim memory of some more harmonious stage of development, poets and philosophers have universally reconstructed the buried traces of a Lost Paradise, a stage of pure experience that haunts the memory of all sensitive men, whether in death, in dreams, or in moments of rare attunement to some forgotten splendor of existence.

# -3-

Such an image, persisting up to death, strong enough to finally replace all other memories, explains the sense of immanent mystery that surrounds our waking consciousness. Once we have perceived this mystery, all that life has to show becomes suggestive of an ultimate secret, the desire for which is tantamount to promise of attainment in the future. This accounts for the traditional obsession of man with states of "otherness" (alterations of consciousness) and his determination to recover by any means the "otherworldly" reflection of his inmost memory. "Otherness," whether induced or gratuitous, and the notion of immortality are both rooted in the archetypal recollection of paradise, since in each of us there operates at once a longing for its return and a psychological assurance of success. Theology generally states that salvation must begin with "re-birth," affirming the innate belief in something that has had a prior existence. In this sense, the idea of eternity itself, at least insofar as we are capable of visualizing it, is fashioned from the half-remembered world that we occasionally perceive in moments of "otherness." It is scarcely surprising that the re-creation of a state wherein the persistent image is momentarily regained should have played an important role throughout human history, since it is in such instants of transport that we are directly assured of the truth of our poetic beliefs.

The employment of drugs to bring about the required alterations of everyday consciousness is an ancient one, as old as men's reverence for prophetic trances, clairvoyance, and related visionary powers.

The Greeks acknowledged wine to have divine attributes, which were revealed by Dionysus, the son of Zeus himself. Through the intoxication of alcohol, knowledge of a transcendent sort was vouchsafed the partaker—hence, the institution of the sacred Dionysian, out of which evolved the art of drama. The Latin proverb "*in vino veritas*" expresses the same veneration for wine's intuitive gifts, gained by transcending the limits of lifeless, analytical thought. Cocaine, which produces hallucinations, was believed by the Incas to be of sacred origin, since it produced even more intense alterations of perception than alcohol, thus finding its way into the religious ceremonies of the Ancient Peruvians. Opium is likewise said to have played a part in the religion of Ceres, the Roman goddess of agricultural fertility. Since time immemorial, the poppy has symbolized for poets and visionaries the gifts of transport and dream. It is also well known that the followers of Hassan-I-Sabba employed cannabis in order to attain glimpses of the Mohammedan paradise, whence was derived the name *hashish*, by which the drug is also known. Whether it be the *ibogaine* of Gabonese Africa, the little-known *Ayahuasca* of the Colombian Amazon, the *ololuiqui* seeds of the Mexican Mazatecs, or the legendary *soma* of the Hindus, chemical means of peering into the contents of the inner mind have been universally prized as divine exordia in man's quest for the beyond, and as such have well deserved the hieratic awe extended to them, before the coarseness of utilitarian minds reduced them to the status of "dope." Duly recognizing the power possessed by such drugs, their employers looked upon them with sacramental veneration. Indeed, their sacred origin alone generally protected them from secular abuse, since the mysteries that they revealed to the Initiate would have otherwise been profaned, as the Peyotists maintain today. As ever, the result of uncontrolled expenditure was diminishing returns, a fact that distinguishes the modern addict from the believer in poetic truth. Significantly enough, the materialist habitué of the present employs his intoxicant or euphoriant as an escape from pragmatic commonplace, rather than as a means to positive experience. Lacking the faith that naive imagination possesses, the exclusively practical mind is aware only of the empirical facts that it sees in a situation. Nothing of the inner meaning that poetic belief creates can be revealed to it. Such a mind is forever denied the experience of "otherness"—hence cannot appreciate the deeper dimensions of man's spiritual life. The human content of experience is subsequently reduced to the level of commonness, and the religious sense, to dreary causality.

Perhaps a new significance can now be attached to the recent discovery of the hallucinogenic alkaloids, one of which I frankly feel has been overlooked by the prosaic pharmacological mind. Medical men have widely heralded the possible uses of mescaline, LSD, or psilocybin in elucidating the etiology of mental disease; researchers hope to shed new light upon the workings of the sick mind with the fact that such drugs can mimic various aspects of schizophrenia, thus suggesting a connection between the subtle chemistry of the body and psychic phenomena. But, the empirical limitations of the medical scientist have caused him to virtually ignore the deeper connection between man's archetypal being and the forms assumed by mental illness, since it must follow that every manifestation of the human mind, whether "normal" or bizarre, is but a reflection of its inner contents. Hence, it would seem that the empirical study of mere histological, neurological, and physiological effects produced by hallucinogenic drugs is much too narrow to reveal anything ultimately significant in this field, any more than the discovery of normal metabolic mechanisms could explain the meaning of ordinary human experience. If what we have suggested is true, it would follow that the psychic revelation of the drug-experience might be investigated to increase our knowledge of the subconscious, and not merely of neural hormones or metabolic disturbances. A very few scientific men have perceived this possibility, including Dr. Karl Menninger, who suggested at a symposium on LSD and mescaline (1955) that the ideological importance of the peyote experience be further investigated and that some scientific analogue of the mystic rituals of Southwestern Indians be sought. Generally speaking, only non-pharmacological men like Dr. Jung or Aldous Huxley have shown any interest for the philosophical aspects of mental behavior because their colleagues are restrained by lack of cultural background from taking imaginative steps not adequately defined by the disciplines of their field. Very few investigators have even the imagination to experiment personally with such drugs or to carry their findings beyond the notation of blood pressure, serotonin level, pupil size, or vague references to "patient anxiety." This is all the more stultifying because it is the subjective activity of the mind alone that reveals the significant patterns underlying human history. Rather than suggest that we ignore the physiological research of pharmacology, I am urging that anthropological and philosophical attention be also given to the revelations of the hallucinogenic experience, since this elemental encounter with the primal memory can provide us cultural evidence of the secrets hidden within the human soul, secrets that may someday help explain the very meaning of human existence.

The line that divides the sane and the psychotic is indeed a tenuous one. That which passes for "normal" in one culture may appear as "abnormal" in another. Religious ecstasy, for example, widely accepted in previous ages, seems to the pragmatic modern world as a symptom of hysteria, although the experience lives on in the curiously split personality of Western tradition. Modern culture being the result of historic processes, we can find many such anomalies manifest at various levels of experience. Since life is never reducible to a simple consistency, the many facets of its meaning can scarcely be integrated into a single attitude or a simple personal relationship with the world. Many of its undertones are anachronistic, yet at the same time emotionally appealing. Hence, the conscious ego faces the problem of ordering its experiences into an arbitrary system of rational and emotional categories, the former being integrated with the outer world and the latter more or less confined to the subconscious, where it nevertheless continues to function in its own imperious way. Insanity, broadly considered, is a breakdown in this integrative process. The individual, as a rationally differentiated organism, maintains a clear definition of its relationship with the outer world. Under emotional stress, the conscious ego may become depersonalized, allowing its primal store of archetypal material to emerge more freely. In such a pre-rational state, the processes of objective equilibrium are relaxed in favor of the subconscious world of the individual, providing us with a means of understanding the nature of the mind itself. Ancient history intuitively understood the value of the insights thus gained into the pre-rational depths of human consciousness. Now that we are capable of synthetically recalling the subjective images that lurk behind the veil of our intelligible world, it would be most unfortunate if proper use were not made of these means of carrying our search for human understanding still deeper into the darkness of man's primordial being.

# -4-

This phenomenon of return to pre-rational existence has many archetypal parallels in human history. Myth and philosophy have universally alluded to the Primal Chaos out of which the forms of being evolve. Our late discoveries are merely restatements of such ancient notions of the Tantric Darma Kaya, the uncreated Eternal, which externalizes itself in the plastic forms of the universe, or of Nietzsche's Dionysian Will, which dreams itself into the Apollonian world of beauty and order. Perhaps these are themselves archetypal memories of man's development from the interior world of childhood to adult objectivity, where dreams must become directed activity. Viewed against the practical necessities of work-a-day reality, this memory appears as an opposite pole of our consciousness, asserting itself in the shape of the mysterious or the uncreated. Since its domain is prior to the rational intellect (although the intellect alone can translate it into art), it can be experienced at its fullest only by re-creating some state of irrationality—hence, the persistence throughout history of the longing for pure, non-conceptual experience that, I believe, explains the Romantic tyranny of chaos over life, of death-wishes over reality. It explains man's perennial desire to transcend the limits of reason, to recapture the state before the encounter with the Tree of Knowledge, before the imposition of rational forms upon purely existent phenomena. Like Goethe's Faust, who descended into primordial chaos to obtain the secret of life, the sensitive being must keep alive the dark, fertile dreams within him, in order to remain creative; otherwise, the balance of his two worlds may swing over to the mere objective and result in sterility.

A possibly new kind of psychoanalysis might also be suggested by the revelations of psychotropic medicine, leading to a more fundamental view of human experience than Freudian psychology has hitherto supplied. The anxiety and rootlessness of modern life will be seen to be the result of inner barrenness, created by the frustration of our subconscious needs in a predominantly rational society. Empty utilitarianism, deprived of poetic faith, forces the sensitive being to reject reality, since reality is devoid of the emotional appeal that it formerly enjoyed. Man is a creature whose necessity is to dream; when the individual, or the race as a whole, is frustrated of its inner vision, it must suffer as when deprived of material health. Human beings are nostalgic for the fulfillment of their primordial Life. Their anxiety is not the result of fear, since modern life is more secure than ever before, but of the loss of inner destiny, of the devitalization of the illusions revealed in the yearnings of myth and poetry. Man's sense of "otherness" has largely been extinguished, along with the power to transcend the naked world. But, even if the individual can no longer look to his old religions for belief, he can be helped to achieve some kind of existential re-integration with life, by being made aware of his forgotten resources, by re-examining the validity of material existence as it appears in the images of his primal memory. Hence, the greatest problem of mental hygiene is to bring modern consciousness once again into contact with the older layers of the human psyche. Life must once more become a ritual encounter with the poetic values of the far off land; the vast archetypal world beyond the threshold of waking consciousness must be reawakened and celebrated anew in the universe about us.

This task of rediscovering our lost eidetic powers is the goal of human wisdom. To transform the work-a-day into what existence signified on the morning of life's innocence is perhaps the ultimate accomplishment in life, for life's greatest offering is life itself, i.e., experience—hence, the more acute our experience, the more real existence will become. During the hallucinogenic intoxication, such intensified perception is generally encountered, for which reason primitive peoples have prized the experience as a periodic process of renewal, restoring in turn the significance of ordinary life. I do not think that this contradicts the aim of modern psychiatry, which is to dispel irrational anxiety through self-understanding, but rather complements it. Life must not be merely rational and free from malfunctioning, but meaningful in itself; it must have intrinsic worth for the mere living of it, and this demands the restoration of the poetic vision that first created our sense of life's inherent values. Those values, I believe, can readily be deduced from the myths and symbols of mankind, from the suggestive patterns of mental illness, and now from the archetypal experiences of the psychotropic hallucination. Hopefully, the immense background of anthropology, literature, comparative religion, and philosophy can someday be brought together with psychotropic knowledge to restore our sense of religious encounter, which coarse utilitarian existence lacks. Reality depends upon the inner meaningfulness of things, not upon their physical proximity; since meaning arises in the heart of man, the way to make the world most vividly real is to re-encounter the values of the heart in the objects of material experience. By making the external world synonymous with the world of primal vision, life can be helped to achieve its primordial fulfillment. No other solution will satisfy the emotional need of man, nor answer the endless questions about the "meaning" of ultimate reality.

# -5-

Religion has been defined as the "finite become infinite" (R. K. Blythe, *Zen in English Literature and Oriental Classics*). Albert Einstein described it as the sense of awe and mystery attaching to our contemplation of life. Both descriptions suggest that pure experience is made more vast by our emotional identification with it, evolving the aura of infinitude that is the essence of real religious feeling. Religion, far from being merely the belief in supernaturalism, is the ultimate dimension perceived in reality, the transformation of the indifferent into the meaningful. No matter what the nature of metaphysical "reality," it can have no intelligible content without final reference to the objects of experience; hence, all concepts of reality, spiritual or material, aim at realization in the objective world. The ultimate meaning of existence is to be found in the glorified forms of our experience, elevated to the religious through sharpened emotional focus. This process is described in William Blake's famous quatrain:

> To see a World in a Grain of Sand
> And a Heaven in a Wild Flower,
> Hold Infinity in the palm of your hand
> And Eternity in an hour . . .
> (*Auguries of Innocence*)

A striking paradox thus presents itself: The deepest spiritual awareness results in the purest sort of affirmative materialism; on the other hand, the greatest folly is abstract materialism devitalized by idealism.

American "materialism" is in actual fact a form of idealism, which seeks to make commerce and gadgetry a spiritual achievement. It avers that life has somehow become more meaningful because of the perfection of our machines. Such an ideal, however, is an ultimate denial of life, for machines can only implement the prior values that life itself possesses. Therefore, to equate life with mechanical progress is to abstract experience into a kind of intellectual propaganda. Material advantages, elevated to mental symbols, become spiritual objectives, insidiously usurping the genuine values of the blood and soil. Ironically, material objects are no longer regarded by the "materialist" as pristine material objects, but are transformed into philosophical, moral, social, or mental values. Experience becomes entangled in concepts, and man is enslaved by a misguided, machine-oriented idealism.

> All we have gained the machine threatens, so long
> As it makes bold to exist in the spirit instead of obeying.
> (Rilke, *The Sonnets to Orpheus*)

The only way to assert our authority over matter is to reject the intellectual slogans of this pseudo-materialism and to return to the original materialism of our primal memory, where the objects of experience stand forth in their own natural light, perceived as sheer, potential existence, as surfaces, textures, shapes, and elemental forms, electrically acute with the mystery of naked Being. D. H. Lawrence described this natural materialism—which can also be detected in Christ's reference to the lilies of the field, in the Buddhism of China and Japan, or in a thousand other poetic guises—as follows:

> It was a vast old religion, greater than anything we know: more starkly and nakedly religious. There is no God, no conception of a god. All is god. But it is not the pantheism we are accustomed to, which expresses itself as "God is everywhere, God is in everything." In the oldest religion, everything was alive, not super naturally but naturally alive. There were only deeper and deeper streams of life, vibrations of life more and more vast. So rocks were alive, but a mountain had a deeper, vaster life than a rock, and it was much harder for a man to bring his spirit, or his energy, into contact with the life of the mountain, and so

draw strength from the mountain, as from a great standing
well of life, than it was to come into contact with the rock.
And he had to put forth a great religious effort. For the
whole life effort of man was to get his life into contact with
the elemental life of the cosmos, mountain-life, cloud-life,
thunder-life, air-life, earth-life, sun-life. To come into
immediate felt contact, and so derive energy, power, and a
dark sort of joy intermediary or mediator, This sheer naked
contact, without an intermediary or mediator, is the root
meaning of religion . . . a terrible cumulative effort . . . to
come at last into naked contact with the very life of the air,
which is the life of the clouds, and so of the rain.
(New Mexico)

Nature has here become a religious experience through some
deep, visceral encounter with matter. Freed from cerebral propaganda,
the nakedness of childlike perception has grasped the inherent
fascination of things in themselves, which is a projection of our own
primal vitality. All Life is sensed to be ultimately related, and the
same deep forces that operate within the blood are reflected in the
sensuous proximity of the objects about us. We must remember that
children possess a strong psycho-sexual curiosity, as Freud pointed
out, and their experience is by no means abstract or idealistic. Our
own appreciation of things is but an extension of this early ability
to be impressed by the sensuousness of life, felt electrically within
the flesh. Having realized the ancient thrill of this primal awareness,
surfaces again become tangibly rich, shapes become manifest beauty,
and natural functions become awesome mysteries. This transfiguring
significance is repeatedly described in mystical literature. The precise
terminology is variable, since the author's philosophical training
may tend toward a "spiritual" interpretation on the one hand or
a materially "concrete" one on the other. But, in either case, the
experiential palpability of the sensation inevitably transforms the
world of objects that we inhabit, as in the case of medieval ecstatics,
who imagined the putrefying excrescence of their wounds to have the
odor of perfume, or in the subjects of Japanese haiku poetry, which
sees transcendent beauty in the commonest scenes:

Beside a well
of foul water
flowers of the plum
(Issa)

This special awareness, which is the essence of all poetry, is unquestionably the most striking feature of the LSD or mescaline experience. Aldous Huxley afterward spoke of the "miracle of naked existence" that presented itself to him, laden with intrinsic, increased significance; yet, the miracle consisted not of some external or ideal quality, but of perfectly contained concreteness. Indian philosophy calls this ultimate reality *tathata*, which may be interpreted as "suchness" or "thatness," indicating that the acuteness of existent forms is such that their own actuality conveys their total meaning. One might reasonably question whether or not some chemical means can arouse a genuine awareness of *tathata*, but if what we have earlier stated is true, our understanding of reality is but a product of all earlier experience and may be recalled from the depths of memory by various means. The problem of reality and the drug experience is not the moral one that Baudelaire raised in his *Les paradis artificiels*, for we are not attempting to obtain godly powers from beyond, but to merely reawaken what we already possess. It avails nothing to raise philosophical objections to the resulting vision, which unquestionably exists, and is afterward preserved in the memory, where "dreams" and "reality" have the same value. Neither is there anything of the dreamlike intangibility in the hallucinogenic experience that we associate with a narcotic intoxication. The material world is never more concrete, nor its "meaning" more clear in the fullest sense of the word; the archetypal splendors of our far-off land are quite unexpectedly recaptured in a blaze of pure presence, beyond which the human imagination has never projected itself.

# -6-

A few moments after ingesting a suitable dose of LSD or mescaline, one suddenly and unexpectedly notes that familiar objects in the room have acquired strange qualities. Without altering their appearance, they begin to suggest new facets of meaning that elude analysis: invisible as electricity, yet irresistibly real. The sense of "otherness" that we earlier described begins to unfold, revealing forgotten glimpses of adventure and mystery. A calm, euphoric tranquility pervades the mind, which suddenly discovers that it is gazing on pure, timeless reality.

Things are no longer fragments of some metaphysical system, but primal objects whose beauty is integral with their essence, as is the blue of the sky or the wetness of water. All the values that have been previously taken for granted are suddenly impressed upon the beholder as palpable notions, such as the quality of straightness inherent in lines, the smoothness of surfaces, or the symmetry of some design. These geometric archetypes, which are the basis of the plastic arts, are revealed with such tangible acuteness that a fresh, vital aspect of matter is disclosed, corresponding to the suggestions of myth and legend. One is overwhelmed by the supreme fact that a wall is flat or that a line is straight; these are no longer abstract categories of geometry and space, but splendid actualities to be contemplated with endless satisfaction.

Colors and textures are even more strikingly revealed; the sheer quality of redness or whiteness seems to carry in itself the ultimate meaning of Being, which is seen as shining existence, above and beyond all theory. While gazing upon the surfaces of an ordinary room, one realizes that the noontide splendor of creation is no less expressed in the humblest aspect of a common rug, wallpaper, or piece of furniture, now transformed into the gold of celestial vision, palpitating with undivided significance, as the fairest petals of a rose or the flesh of a beloved. The richness of these preternatural hues assumes a quality of physical density; the idea of color is no longer a function of apperception, but a concrete substance, which can be touched and felt. The simplest texture is suddenly drenched with the kind of poetic significance that has always tantalized the dreams of our greatest painters. One gazes upon a newly discovered world of wood-grains, fabrics, lacquers, glazes, and fragile transparencies all cloudless and pellucid as the beauties of Paradise, when first beheld in childlike wonder. The remains of food on a soiled plate are more miraculous than the colors of a Van Gogh masterpiece, and the mystery of a colored button lying on a white tile drain surpasses the whole Arabian Nights.

A certain Colonel Blood, whose visionary experience under ether is recorded by William James, summed up his enlightenment as follows: "The universe has no opposite." As meaningless as this drug-inspired sentence might seem, it actually expresses the very "suchness" that is described in Buddhist teaching. For there are no symbols in the psychotropic vision; reality is pure, illimitable, and non-contingent; and only ideas have their contradictions, whereas actuality simply is. As the acuteness of this psychotropic perception increases, conceptual experience diminishes, allowing the necessary conditions for experiencing this incomprehensible simplicity to unfold of themselves to the extent that the rational ego is inhibited; that state of childlike understanding which Christ likened unto Heaven remains, confronted with the primal objects of man's lost Paradise. It is thus that the inner light of the mind is able to perceive once more what memory compels it to recognize in its transient existence; out of the far-off land there begins to shine the distant reflection of infinity, reawakened from the rudiments of primal experience, emanating as it were from the inmost heart of transfigured matter.

## -7-

Since the dawn of history, this visionary process has been attended with supernatural light; almost without exception, religious experiences of the supra-mundane sort are accompanied by brilliant luminescence, such as the English mystic, Henry Vaughan describes in his poem, "The World:"

> I saw Eternity the other night
> Like a great Ring of pure and endless light
> All calm, as it was bright . . .

Sun worship, the endless day of the Apocalypse, the veneration of jewels and bright metals, the flaming rose of the Unio Mystica, or the overpowering light experienced by Mohammed and St. Paul, all have their basis in some mysterious process, which is now set in action by the hallucinogenic drug. The colors that have already been seen in their Eden-like purity begin to glow like subtle neon, pulsing with intrinsic inner brightness; a Kodachrome-spectacle of intense brilliance transforms the humblest shapes of experience into vessels of Heracletian fire. The world of mythology and ecstatic vision suddenly opens up to the perceiver, who begins to experience for himself what history has restricted to a fabled few. Man's instinctive awe before the mystery of light is reawakened by the spectacle of cosmic luminescence, noumenously shining in every surface or exploding like millions of Milky Ways in his field of vision.

---

Our identification with the miracle of consciousness has long been symbolized by the phenomenon of light. Poetry and language abound in metaphors such as "illumination" and "enlightenment" when referring to higher forms of knowledge, while darkness is always associated with chaos, negativism, and death. Goethe's dying words, "More light!" were the last request of one who instinctively clung to the vanishing embers of existence. Not only our respect for the sun, which is the source of our life, but also the peculiar organization of our brain, which invariably provides the "supernatural" brilliance seen in extreme moments of "otherness", accounts for the traditional association between luminescence and the Divine. From pre-Socratic Greece to modern physics, with its equation of matter and energy, philosophers have repeatedly ascribed characteristics of fire to Ultimate Reality. Tibetan texts on Tantra claim that Being is essentially a clear, timeless Light, in which waves of mentation create patterns of phenomenal perception. To see this perfect Light involves a mystical discipline not unrelated to our experience of "otherness", whether initiated by esoteric training or the pre-rational mechanism of psychotropic drugs. Indeed, Hindu mythology alludes to such a legendary substance that enabled the partaker to enter into the Vedic world of Ultimate Light:

> We've quaffed the soma bright,
> And are immortal grown;
> We've entered into light,
> And all the gods have known.
> (*Rig Veda*)

It is not surprising that the Indo-European races as a whole identified their gods with luminosity. The Indo-European root *dijeus* referred to the sky, from whence was derived the Sanskrit noun *dyauh*; this term, combined with the word for "father", gave the Vedic name of *Pita dyauh*, the "Sky-Father", who was also cognate with the Greek father of the gods, *Zeu pater* (Zeus). In Latin, his name was *Juppiter*, similarly derived from *dies pater*, ("day" or "sky-father"). All of these related words express the same ineffable longing that human imagination has connected with the light of day, especially when it is seen at sundown, disappearing in the distance, as a final reflection of the far-off land. According to the legends of many races, Eternity lay in the West, where the eye could no longer follow, as it watched the last faint colors of day recede into darkness. The effect upon the poetic imagination of this last brilliance, like fire behind the already black silhouette of some elevated horizon, can easily be appreciated. Yet, through the means of psychotropic perception, this experience may be expanded to a level of transcendent sublimity. I have thus watched the disappearing light of evening, spread out as a rosy mist upon the mysterious mountains and distances of my own desert. Perched on some high hill, it seemed to me that the magic of that realm, which we see symbolized in light, was itself a glimpse of the Unattainable, transforming material space into a suggestion of the Infinite. This sum of all beyonds, transparent and luminous, like a remote prospect of harmony or a recollection of ancient paradise, carried the mind to the limits of its conceptual powers, as if the very secret of the universe were to be perceived in the mystery of those last gleams, returning to the ultimate Home of all things. This utmost nostalgia, coupled with fear of the oncoming darkness, must have compelled our ancestors to recognize the reflection of their own interior vision in the unfathomable light that they daily beheld, withdrawing by evening to its primal source and again reborn each morning. Indeed, the whole archetypal meaning of the "otherworldly" seems to reveal itself in those magic distances, illuminated by the departing rays of the sun.

Yet, how miraculous to behold this same light during moments of visionary attunement, pouring forth in the inner eye of the soul! All of those who have encountered the "otherworldly" in mystical or morbid states have spoken of the brilliance in their optical field; the identification of one's own vital force with the external light of infinity would be a scarcely illogical reaction to this similarity between microcosm and macrocosm. In fact, with the help of LSD and mescaline, one may actually re-experience this metaphysical insight—not intellectually, but directly, as it must have happened in the intuitive subconsciousness of the race. The archetypal basis for this universal notion becomes accordingly apparent from the information released by our cerebral mechanism, revealing the identical sources of our memory and the haunting suggestions perceived in nature. As in the sun itself, this light of interior vision is a world of burnings, of seething fire and creative force, flowing incessantly from one luminescent form into the next. With eyes closed, this inner fire becomes a series of kaleidoscopic scintillations, varying from geometric forms of surpassing loveliness to images of utmost sublimity, all perfused with uniform brightness, reminding one of the solid gold backgrounds of Byzantine art, which attempted to portray the kingdom of God with shining metal. In the deeper states of hallucination, matter itself becomes a play of pure energy, curiously identical with the view of modern physics. The nervous system actually feels this electric force feeding the forms of perception, congealing into substance, and then reconverting itself into a sort of luminescent *élan vitale*. Visionaries have experienced this equivalence of matter and energy since ancient times, as recorded, for example, by Zen adepts during the state of *satori*:

> The "God of Matter" has just taken off his mask of glacial immobility, and, behold, is transformed into a prodigiously moving, fluid, impalpable energy. His countenance, which once appeared sombre and dull, is now lit up with ever more dazzling clarity. The silent fairyland of light, perpetually unfolding in the heart of the smallest grain of sand, far exceeds in splendor the most brilliant fireworks we could ever hope to see.
> (Robert Linssen, *Living Zen*)

Modern scientists will dismiss this vision as an illusory coincidence; yet, they would do well to remember that the contents of the human mind have remained consistent throughout history, being merely adapted to the language and conceptual methods of various ages. The present results of the experimental laboratory are only meaningful according to the subjective framework of our empirical age. It is still the "hypothesis", representing the accumulated imagination of the race, which guides even material science, since the experimental method has nothing to work with until marshaled into the service of some subjective idea. If its practitioners prefer to smile at the mystic who arrives at certain identical notions, they should reflect that physics today has merely modified the insights of ancient times to its own subjective understanding, and atomistic ideas, coupled with theories of vital force, have enjoyed an uninterrupted vogue right up to our own time. Each successive age has recast them according to the methods of reason it employs. It would seem, then, that the primal ideas themselves, with their organic significance in the human mind, are more deserving of attention than the transient terminologies used to frame them. Indeed, one might seriously question whether or not modern physics would have ever conceived of the forms it employs in describing "reality" had they not existed as a cultural heritage all along.

Only the ignorant are unaware of how great a role the past plays in their every thought. Perhaps the greatest significance of the deeper stages of mescaline or LSD perception is that the subject now stands face to face with this integrative bedrock of memory. Even more meaningful than the brilliant visual phantasmagoria are the ideational processes that they reveal, processes that are in fact the basic content of human experience. Unlocked by some mysterious power of the drug, an unbelievable profusion of ancient symbols pours forth. Ideas underlying the oldest myths appear with eidetic clarity; the spectacle of light and motion becomes an absolute experience of the Oneness of being; pantheism and indwelling Divine Energy may alternate with a spectacle of the atomic force pulsing through one's body, projecting itself through the vastness of space into shining galaxies. The Eternal Flux is succeeded by the changeless majesty of Platonic Ideas; one is no longer bound by space and time, but senses his identity with Life itself, perceived as a timeless moment, without beginning or end. Relativity, the Harmony of the Spheres, of the Infinite, of the Infinite are seen, not as their external representations, but as the ageless impulses that gave them birth. A descent into the womb draws one back onto the primal slime of creation, glistening and palpitating with visceral wetness like a red, cavernous swamp. Hosts of sperm-like animals swim upward into the translucent glow of some prehistoric springtime, aglow with the force of creation, oozing up through the roots and trunks of rank vegetation. Legendary creatures peer out of the jungles of unconscious memory; sexual phantasmagoria dissolves into mythical spectacles of reincarnation, driving the Round of Existence with an electric rhythm of immense carnal lewdness. Human bodies tumble past, emanating their sheer suggestion of blood and instinct, charged with the fullness of every poetic feeling that man has experienced since the origin of his race. Love, exaltation, joy, and suffering pour out of the well of time in images of shining clarity, succeeding one another as threads in a living tapestry, wherein are woven the patterns of primordial meaning that form the heart of human consciousness. Within this iridescent sea of archetypal pictures, the secrets of the *mandala*, the Rebirth from water, the phallic serpent, the Eternal Feminine, the uterine caves of the Venusberg, and the beasts that appear to us in nightmare, all unfold in their turn. Whatever mysteries man has projected into the wandering clouds, the heart of the atom, or the wisdom of philosophy await discovery in the depths of the soul that gave them birth. This is the ultimate exploration that can ever be undertaken.

In addition to these collective and archetypal memories, which should prove of immense value to anthropology and philosophy, the personal past of the individual returns with amazing total recall. It is this particular phenomenon that currently makes LSD of use to psychiatrists, who experimentally employ the drug as an analytical aid in uncovering suppressed material from the subconscious mind. The occasionally claimed validity of memories prior to birth must be questioned, since the embryo, even if possessed of the ability to record its experiences in the womb, would certainly not have formed the conceptual mechanisms that adult memory attributes to these prenatal events. Yet, the unexplained clarity of long-forgotten scenes from very early childhood raises the question whether or not some primitive mental process indeed might be active in the fetal consciousness, which later stimulates the universal occurrence of uterine and birth fantasies in the dreams or hallucinations of various races. I am disinclined to believe that such fantasies can be direct memories; nevertheless, the continuity of human consciousness from the present moment back to an undetermined point in the past suggests that there may well exist some psychic record of a diffuse sort, beginning even before the neonatal state.

Imagination certainly plays its part in these fantasies, as is proven by the immense creative power of the mind under hallucinogenic stimulation. Daydreaming can lead from one arbitrary picture to another, devised quite at random and under the complete control of the will. Yet, the eidetic power of the memory is at the same time so strong that actual events are recaptured with the concreteness of direct experience. Tastes, odors, minute visual details, and attending emotions are present in perfect clarity, as fresh as at the moment of their actual occurrence. This absolute ability to transport oneself back into the scenes of former life mimics possession of supernatural power, which explains why certain peoples even today employ hallucinogenic drugs in practicing clairvoyance and communication with the spirit world.

The basic anthropological connection between memory, the world of the beyond, and the existence of ghosts becomes obvious through the juxtaposition of past and present, which LSD or mescaline projects upon the screen of inner vision, where a thousand scenes from bygone years appear as if the movement of time were indeed an illusion. What imagination attributes to the state of innocence, the intellect can examine directly as a re-established fact, here and now. To actually relive a phase of life wherein the clouds, the green boughs of secret trees, the scent of ripening apples, or the cry of a distant bird seems more important than the margin of profit shown in dusty bankbooks is not merely an esthetic experience, but direct evidence for our hopes of a simpler, more golden-rich existence amongst the basic things of blood and earth. For some years I had in my papers a youthful attempt to record the almost visionary impression that a summer afternoon on Clapham Common once made upon my boyish mind. From the viewpoint of practical reason, such powers of perception seemed more romantic than real, yet a capsule of mescaline one day restored not only the memory of that vanished poetry, but also the tangible experience itself, which irrefutably transcended any preconceived imagination. Such gratuitous return of lost energies, even temporarily, is like the sight of a distant beacon, redirecting the conscious mind toward the image of reality slumbering within. Our unexplained dreams and preferences we thus understand to be reflections of our far-off land. Amongst the symbols and archetypes of cultural life, the entirety of our past lies root-active, waiting through the years of intellectual distraction to be reawakened in adult experience. Buried in this store of memory are the patterns of all the perceptions that we experience as reality; for this very reason, the psychogenic reencounter with our primal vitality can reveal the meaning of whatever ordinary existence has to offer of genuine significance and fulfillment of inner destiny.

# -8-

M any of the symbols of various cultures can be traced back to the nature of the childhood experience, stored in the primal memory of the race. Language betrays the universal idealization of the past ("innocent as a child", "the young at heart", "sleeping like a babe", etc.). Literature and legend record the endless obsession of mankind with eternal youth, the poetic perceptions of childhood, and the open-eyed appreciation of youth for the mystery of life; the motifs of the Wandering Jew, the aged Faust who sells his soul for recaptured vitality, the Hesperidins Isles of perpetual springtime, the ageless figures of occidental and eastern mythology, and the endless longing of poetry for the youthful prelude to life, all show that existence is enjoyed more in retrospect than in actual fulfillment. Even our dreams of the future are an optimistic appraisal of our ability to someday construct the ideal situation that memory suggests. The adult necessities of the present are viewed against the idea, real or imaginary, that we once possessed gifts that are temporarily submerged underneath the burdens of responsibility:

> And yet I know, were'er I go
> That there hath past away a glory from the earth . . .
> (Wordsworth, *Intimations of Immortal*)

Human existence is experienced as a widening circle, whose interior is the accumulated past and whose periphery is the line that we call the present. As it expands, reality is stored in the memory, and the tasks ahead are seen as challenges to our continuing labor. Time thus begins to exercise its tyranny over life; the painful consciousness that the present cannot be permanently secured, any more than quicksilver running through the fingers, generates the disease of human anxiety. Past memories mock the possibility of future failures. In learning to savor memories, we are inevitably confronted with the problem of maintaining their imagined intensity in the face of fleeting powers. The need to continually test these powers creates a kind of self-consciousness that is tantamount to psychic onanism, or the obsession with self, and the need to continually sustain the enjoyment of one's own existence.

This anxiousness regarding life's imagined limitations is an outgrowth of an exaggerated ego-differentiation process, which gradually destroys life's spontaneity through self-consciousness. This explains the profound significance of the Biblical dictum: "For whosoever will save his life shall lose it". For, in worrying over much about the self, one exhausts one's vital energy and inhibits all normal activity. True "selflessness" is not to be construed as mere maudlin "charity" or the so-called "love of others" (which is more often than not a piously concealed means of glorifying the self); it is rather the complete forgetfulness of self-consciousness and its accompanying anxiety, which permits one to act with natural innocence and efficiency.

The goal of true wisdom is to act in harmony with nature and develop one's gifts without compulsion or anxiety. This sort of natural life is what Taoist writings call *wu wei*, or "non-striving action", which is much like the swimmer who allows the current to bear him to his destination, rather than fighting it to his own destruction. But, the intellect, as an ego-differentiating mechanism, will ever rationalize its own motives, defending its own interpretation of reality against the inexorabilities of the world; the childlike mind, on the other hand, which as yet makes no distinction between the self and the world, senses only pure existence, which is undisturbed by mental preconception. This freedom from psychological dishonesty was held by Socrates to be a necessary condition for true knowledge and virtue. Memory and mental habits, he taught, interfere with direct insight because through them the ego projects itself upon the world and is once again entangled in its own pretensions and motives. The more direct knowledge is, the freer it is from egoistic coloration and distortion.

One might question here whether or not the childlike mentality is actually free of egotism; it may be objected that the child seeks only to satisfy its own needs, free of concern for others. We may hold, nevertheless, that without some definite awareness of the relationship between the ego and the outer world, there can be no true sense of self, since there is no definition of the distinction that places self above one's fellow man. This simple existence is still innocent of the motives that create the artifices of individual and selfhood. Both Christ and Buddha showed that bondage to such relationships destroyed such selflessness, by creating attachments to the objects of one's own ego:

> Love not anything;
> Hate and envy arise from this same love.
> He who loves nothing hates nothing,
> Is free from all evil bonds.
> (*Dhammapada*)
> If any man come to me and hate not
> his father and mother and wife and
> children and brethren and sisters,
> yea, and his own life also,
> he cannot be my disciple.
> (Luke 14:26)

The distinction between "thou" and "me", sustained by the frail yoke of human desire, pitilessly outlines the nakedness of the self, tortured by its own anxieties, entangled in the subtleties of its own rationalizing. The opposite ideal may be difficult or impossible to attain, yet it necessarily persists in the memory, from whence it finds perennial expression in the teachings of mystics and philosophers, pointing back to the original state before the rational intellect and its self-consciousness. It is most suggestive that the word "rationalize" is derived from the Latin *ratio* (intellect) since that is exactly the function of the mind, i.e., to reckon with its own assumptions and produce agreement between "reality" and its own mode of understanding. Since these assumptions are preserved as mental habits, which subordinate the present to the past, they become entangled in the tyranny of time, thereby contrasting the image of the anxious self with the insufficiencies of its ephemeral existence.

During the hallucinogenic experience, one is frequently obliged to undergo such an encounter with the naked soul, robbed of its "outer-directed" pretensions and driven by the need to rationally cope with the material released from the inner consciousness. The nature of this encounter necessarily varies with the subject, but since one's visions are but projections of the self, the self is inevitably forced to evaluate its own image, resulting in varying degrees of apprehension. Anxieties, fears, practiced deceits, and neurotic habits, all emerge under a powerful magnifying lens, along with the illusions that constitute one's appraisal of reality. To be brought face to face with one's own defects may be a terrifying experience, but the truthfulness of LSD and mescaline is such that it does not spare the beholder unpleasant facts regarding himself.

During subsequent experiments with the drugs, one notes a growing tendency to anticipate their visionary content; these preconceptions result in a sort of spurious experience that is partly an attempt to rationalize involvement with some *a priori* anxiety, similar to that which drives us into the sangsaric contradictions of life described by Eastern religions. The first hallucinogenic experience is free from such anticipation, however, and is always a genuine adventure, leading to unknown explorations:

> Mais les vrais voyageurs sont ceux-la seuls
> qui partent pour partir . . . (but the true travelers
> are those who go just to be going . . .)
> (Baudelaire, *Le Voyage*)

But, invariably, the sophistication of advancing knowledge begins to gnaw at the innocent pleasure that one obtains from his view of paradise. One begins to sense that he is involved not with the splendors of the primal world, but with the shadows and souvenirs of his own will, carrying him around in an aimless circle of self-deceit.

Repeated dosages create an obsession with result. The victim becomes anxious lest each experiment will not be as rewarding as the last. He learns by object lesson that he is in search of the essentially unattainable, for the value of an illusion lies in its very unreachableness. He learns the masturbatory pleasure of intensifying desire by means of self-imposed obstructions to its realization. For him, Rimbaud's complaint becomes an obsession: "Real life is elsewhere!" Reality is not here; it is an image burled in the imagination, clothed in romantic symbols of death, the ineffable, and the beyond. The drug-taker or the poet who seeks a temporary taste of the infinite in his inverted world of self-indulgence earns but fleeting freedom from banal reality. The very strain of the deception inevitably collapses in anxiety and disillusionment.

This tragedy of Weltschmerz derives from the fact that knowledge destroys illusion, and such a reaction is irreversible. Once attained, knowledge cannot be gotten rid of. What charm life possesses is beyond logic, and thus incommunicable. Every great man undergoes a purgatorial descent into himself, only to learn that what can be told is not the truth and that truth is what cannot be expressed. Such a confession is not an account of knowledge, but of the paradox encountered in the nakedness of his soul. For that which we understand has lost its charm, like toys that have been dismantled to reveal their secrets. This progressive exposure of life's basic paradox leaves us alone and uncomforted; the sacredness of the primal world, with its authority and tradition, survives only as an unconscious search for our archetypal symbols, enigmatically pointing to some lost world. Doubt destroys our capacities to experience beauty; soon we are able to love, to feel, and to express ourselves. Anxiety, lack of meaning, and the sense of rootlessness increase our impotence, so that the fear of not finding pleasure obliterates our ability to do so. Life's meaning is reduced by the loss of spontaneous vision; finally, the bright colors of paradise are faded into dull shades of grey, and the light of perception is hidden in shadow

This schizophrenic interlude during the psychogenic experience plunges one into absolute limbo of zero-nothingness. Where the casements of eternity once shone with unbelievable splendor, one now sees only cold, dead matter, totally divested of significance, neither joyful nor tragic, but characterized by complete lack of emotion. No one who has experienced this hollow world of vacuity can fail to understand the lusterless existence of schizophrenic withdrawal or its genesis in the cosmic dilemma. Robbed of its illusions, the naked soul stands dead and alone. Hallucination may continue, but the external world is utterly empty. Denuded of their authority, beyond all contact with one's exhausted energies, the earth's primordial beauties are fled; its physical presence becomes sepulcher. Here, in complete negativity, not even pain is possible; having once reached the bottom of the psychic abyss, the mind has spanned the full circle from sheer affirmation to limitless nothing.

# -9-

Perhaps the most remarkable property of mescaline and LSD is their ability to compress into an afternoon the result of many years' experience. Our entire development may be retraced in the retrospect of a few hours; indeed, one may witness the dramatization in a period of minutes of certain processes of racial consciousness, which have extended over millennia.

Medieval mystics often spoke of the Dark Night of the Soul, which preceded the entry into the *Unio Mystica*. This climax to a long struggle with the self was a period of utter despair that finally destroyed the tyrannical ego through its own torment. When the pain of protracted suffering completely disrupted the mechanism of anxiety, there followed the blessed miracle of simple resignation; for the first time in years, the cleansed soul gazed upon primal reality, shining again with its own pristine radiance, free once more from the painful reflection of the exaggerated self.

It was only after passing several times through the hallucinogenic allegory of the human purgatorial, and reawakening into the bliss of a newborn world, that I realized the obvious similarity between this release from bondage and the psychology of Nirvana. As understood by the West, the "extinguishing of the flame" generally implies a negativistic escape from earthly things, but as understood by Mahayana Buddhism, especially those schools that flourished in the animistic soil of Japan and China, Nirvana is the ultimate glorification of the natural and the concrete. We have already referred to certain differences between nirvana and sangsaric life, which emphasized the desirability of freedom from egoistic anxiety; we must now point out the profound connections between this ancient wisdom of human experience and the state presently unfolding in our symbolic drama of spiritual evolution.

The Chinese philosopher, Ch'ing Yuan, once stated that we first see mountains as mountains and waters as waters; with increasing sophistication, we decide that mountains are only perceptual images, and waters, only fleeting, sensory illusions. With full enlightenment, however, mountains are once again mountains, and waters, just waters, for "all manifestations and feelings are identical with the essence of the eternal mind" (*Do-ha-mahamudra*). In the same manner, we have gone from simple perception to the complexities of intellectual sophistry; we have learned the fatality of the rationalizing ego and its anxious search for what is essentially unattainable, i.e., the purely mentative values that it seeks to find in elemental existence. We have fallen deeper and deeper into the sangsaric state of self-consciousness, until by painful necessity, the veil of intricate cerebration has been rent, and the wisdom of simplicity has again illuminated our despair.

This is the original paradise described by Dante, in *la Divine Commedia*:

> The first age was fair as gold; it made acorns savory with
> hunger, and every stream nectar with thirst.
> (Purgatorio, Canto XXII, 148)

The Zen *Saikontan* says: "When you are hungry, eat rice; when you are weary, sleep." Huang-Po adds: "The foolish man eschews phenomena, but not mentation; the wise man eschews mentation, but not phenomena." Is this any different than what Christ tells us in the Sermon on the Mount?

> Therefore I say unto you, Take no thought for your life,
> what ye shall eat, or what ye shall drink; nor yet for your
> body, what ye shall put on. Is not the life more than meat,
> and the body more than raiment? (Matthew 6:25)

Life transcends all search for a meaning, since experience alone is the measure of existence. The highest wisdom, therefore, teaches us not to despise phenomenality, but to seek the infinite in the finite of every moment, for Life is no "problem to be solved, but a reality to be experienced" (G.F. Main). Having learned the futility of an errant search for values beyond life, we have perforce returned to the bosom of earth, but enriched with the darker wine that is total experience, deepened by the intuition that even human sorrow is an expression of the cosmic drama enfolding our frail existence. Desire and pain are no less realities within the entirety of Being than the perfect harmony that it reveals to the vantage point of eternity. Hence we eat, we copulate, and we enact the drama that is Life, again aware of human individuality, but transcending its limited meaning, for each separate self represents the unimpeded interdiffusion of Absolute Reality.

What that Reality represents, revealed in common, individual life, is the highest sort of emotional fulfillment that earthbound man can attain. For me, this final phase of the hallucinogenic experience will always be the most significant; were I to require poetic expression to describe my sentiments at any time during the experiment, it would be here. Yet, I have been able only to gaze about me, beholding the common scenes of my daily life with uttermost satisfaction, transfigured by the glory of a new existence shining brilliantly in the old.

> The old furniture glowed with a miraculous polish, the carpets and curtains were as if new again, daylight three times more brilliant than natural day came in through the windows and the doors and in the air there was a freshness and perfume like the first day of spring . . .
> (Gerard de Nerval, *Aurelia*)

What was merely observed, during the early phases of the intoxication is now felt, elevating knowledge to genuine mystical experience. What this emotion is like exceeds the limitation of language, for inner recognition alone comprehends the meaning of such harmony, which is indeed the "peace that surpasses all understanding." Pleasure is elevated to joy, and beauty, to the sublime; where formerly the world was "divine", it is now perfect presence.

Between those two aspects of a single fact, there lies an immense gulf of mental effort; to comprehend the proximity of the Beyond is exquisite, but to hold it, all in all, is too miraculously simple to be coldly rationalized. One can only touch what is real; one can feel it, one can live it, but one cannot retain it in an empty vacuum of words.

Music, of all the human arts, is alone capable of expressing the ineffable. Since ancient times, man has perceived the Harmony of the Spheres in the ecstasy that the concourse of sounds provokes in his soul. Everything that has been left unsaid is awakened by music, even that at which poetry can only hint. But, the traditions of music relate to music alone; no shapes in nature determine its character. It may become highly formal; but, its habits are its own, and its essence, the feeling that gave it birth—hence, it has spontaneously existed wherever there has been emotion to invoke an audible response.

At the uttermost limit of the LSD of mescaline experience, nothing else can adequately reflect the transcendencey of the mind. During early experiments, I repeatedly listened to portions of *Tristan und Isolde*, seeking in that supreme sound of longing the symbols that would express the nature of my emotions. Wagner's music of transfigured desire has always seemed to me to epitomize the nostalgia we feel for the far-off land. In romantic bliss, one breathes "the world-breath's billowing All" and experiences that "highest joy" that attends our sense of metaphysical release. Yet, later on, the world of Tristan, and its otherworldly ecstasy, had somehow failed to convey the meaning implicit in the final stage of hallucinogenic vision, for only the pure light of day, unlike the dark glow of Manichean redemption, could represent the earth in its reborn splendor. True enough, no other music could so completely express the unutterable sight of those Lost Isles in the far fields of evening, but the mind inevitably returns to the music of Mozart for the springtide affirmation of its total awareness. "*Es ist das hochste der Gefuhle* ... (It is the highest of all feelings . . .)," Papageno and Papagena sing in *The Magic Flute* of their innocent pleasure in each other's presence, of the human analogy between Creation and the bearing of children, sensed in the visceral joy of Life, which is beyond all theory or knowledge. Schikaneder's libretto certainly intended to satirize the earthy simplicity of the bird-catcher and his sweetheart, but the superior insight of the composer transformed their commonplace into a symbol of Infinity itself, quite as compelling in its own way as the "higher striving" of the opera's loftier protagonists. The sound of the female voice becomes laden with all the warmth of human feeling, physically palpable with the burden of sexual richness, in which the inner secret of our existence reveals itself. That supreme mystery, which trembles in the hungry womb or in the delicate fibers of the brain, is transformed into sheer harmony, pouring forth with unadulterated bliss in face of all the tragedies that fate might place in its path. Indeed, the magic of Mozart lies in the quality of affirmation that he turned into art, encompassing even the sufferings that continuously oppressed him. It is not the music of simple happiness, much less of that often described "gaiety", but of serenity, of a harmony beyond all opposites, yet gratefully manifest within them.

One listens now to the sensuous presence of the human voice, with its blood-warmth and its lust for being; one hears the oily richness of instruments, dropping their clots of honeyed loveliness upon the ear. Where habit previously elucidated sounds in context only, the undivided presence of a single harmony now discloses the very meaning of musical consonance. How might one express this exquisite significance of tonic and dominant, suddenly commensurate with the Foundations of the Universe, burning the soul with its unspeakable beauty? Naked sound, with its own perfect meaning, is the final secret of musical art, resonating within the bowels in acute sympathy with the entire physical cosmos. "When I see, I hear, and when I hear, I see", runs a Japanese *koan*; this synaesthetic correlation of mental perceptions suggests the conceptual basis for our persistent belief in the "oneness" of existence. Music thus falls "like drops of golden light" (Hesse) or erupts as a "high tree in the ear" (Rilke). Every mood, every inflection of sound is reflected in visual sensations; the fragile tones, in showers of delicate crystal; and the deeper ones, in an ooze of creamy opalescence. With mounting sublimity, the inner vision beholds the majestic dance of life, swirling in a boundless rhythm of creative activity, flowing endlessly into the forms of eternal Harmony. Every transition is a breathing in-and-out of the world's organism, fulfilling, in its movement between heaven and earth, the innermost destiny of conscious existence. Every sigh, every phrase reveals in its turn the mystery of pain and bliss, infinitely pulsating throughout a universe of sentient awareness. The vast spectacle, with its endless images in musical concourse, recedes into gulfs of widening grandeur, enfolding within its heart the glowing shapes of phenomenality, like jewels upon a net of intertwining creation, perfused with a single identity of sensate bliss. In its very exhalation, the mind discerns its own reflection in the dazzling motion of each atom, trembling before the mighty instrument of the cosmos, enveloping each conscious fiber in a final *Unio Mystica* of incandescent sound. Near and far, the casements of Eternity blaze with resplendent light, unutterably pure and serene, transfigured into a limitless scene of ineffable peace, unspeakable felt in the perfect knowledge of immediate experience. In these ultimate moments of human transcendence, the inner vision is opened to the last of life's secrets, comprehending in our fleeting evanescence the exquisite longing of all existence, lyrically palpable with an unbearable sublimity that transforms our drama of life and death, of love and desire, into distant music, immutably joyful and sad, as if the deepest happiness and the uttermost suffering were one common ecstasy.

This scene of final illumination, perhaps most fully experienced through the symbol of music, is the consummation of the hallucinogenic experience. Beyond this point, its images become paler; its spectacle, less substantial. But, the sense of infinite encounter persists another hour or two, revealing a newborn world outside, sharply etched in the waning light of evening, its lawns and houses calmly resting beneath a transparent sky, indescribably beautiful in their very commonness: the grey pavements, deep in texture, and the pink colors on the mountains, refreshing as iced fruits. Human pleasures become suddenly enchanting; the earth is tender toward its creatures, as if spring were again returning to the heart.

Yet, the total effect of the experience must continue to evade description.

How shall we describe the lofty joy of the mind, or its utter tranquility and satisfaction? Quite unlike the euphoria of stimulants, the effect of LSD or mescaline is spiritually quickening; one emerges with feelings of deepest happiness, with the sense of uplift and ritual encounter. For some, the drug bears curses in place of illumination; yet, the power to bestow mystical vision upon the beholder has been stipulated since time immemorial. The profound pleasure of that vision, the lofty splendor of its affirmation will remain the goal of poets, who must ever search the symbols of their other worlds for words to frame it or for reflections of the Beyond with which to comprehend it. For me, it has been the supreme religious experience, uniting the Infinite illimitably with earth, creating a Garden of Eden for the period of its duration, and revealing in the shapes of ordinary life the Harmony of total existence.

# -10-

Having completed our subjective observation of the hallucinogenic experience, we must ask what the general value of such an encounter might be and, especially, what could be gained by its repetition.

The scientific and philosophical possibilities of the experience will be immediately obvious. We have already pointed out what anthropology might learn from the deeper layers of the human mind; comparative methods will continue to furnish background for such a study, but the mind itself can now provide the original material. Much has recently been done to correlate psychology, anthropology, and religious knowledge, so that a broader picture of our archetypal world may be perceived. With this added means of unlocking the primal memory at will, the basic patterns of human intelligence can be directly observed and, best of all, intuitively understood. One need no longer only imagine the relationships behind empirical experience and conceptual knowledge, for the psychomimetic drug projects them into vivid geometry, into synaesthesic pictures of sensory perception, or into symbolic re-enactments of the categories of time and space. Since our cogitative mechanism is derived from physical experience, it is ultimately understood through physical representations. One sees, for example, that arithmetic functions and sets of objects are actually identical, and not merely equivalent in some mental fashion. Observing reality as a form of "thatness" without cerebral opposition, the mind under mescaline or LSD requires no mental equivalence between objects and "understanding" because it perceives both the substance and the idea of its perception as a direct unity. These tangible images provide the philosopher with a means of intuiting the epistemology of time, mathematics, spatial dimension, and even consciousness itself. Questions of a metaphysical nature are concretely reduced to the experiences that created and then revealed the conceptual processes behind our drama of reality and meaning with eidetic clarity. These problems of human knowledge help us to understand the unity of our sensory world, which is ordinarily distinguished at disparate levels of existence, due to the need for reason to order its perceptions according to the cinematographic principle elucidated by Bergson.

We must take exception to Aldous Huxley, when he states (in *Heaven and Hell*), "Almost never does the visionary see anything that reminds him of his own past. He is not remembering scenes, persons, or objects, and he is not inventing them; he is looking at new creation." Certainly, some of the bizarre caricatures of the greatly stimulated imagination are unlike anything ordinarily recognizable, yet repetition of the experiment will reveal how basically related these visions are to human experience as a whole. In fact, one learns the relationship not only between these images and previous memories, but also between their symbolic content and the very mechanism of thought itself. What one experiences under LSD or mescaline is a product of his own conceptual imagination entirely; no one has yet reported seeing a new color, previously unknown, or hearing a sound, the elements of which have never before been encountered. Again, we must disagree with him, when he writes, "Here, at the limits of the visionary world, we are confronted by facts which, like the facts of external nature, are independent of man, both individually and collectively, and exist in their own right." The forms beheld in these "far-off regions" are said to "serve as intermediaries between man and the Clear Light". Such a statement, however, assumes *a priori* some sort of metaphysical existence beyond the present world, such as the *Bardo Thodol* suggests, of which human perception is but a secondary reflection. Whether or not this existence should prove to be essentially spiritual, as the mystics have suggested, or purely material, as nineteenth-century science believed, our notion of such a state is completely limited to the forms of our conceptual understanding—hence, it seems most accurate to comprehend our earthbound intimations of an "otherworld" as a vague recollection of a former manner of experience that we have since been forced to abandon. Indeed, the nature of human knowledge makes it unlikely that we shall ever approach "metaphysical" truths through existential experience of any kind.

The antagonism between "matter" and "spirit" seems, in light of the hallucinogenic experience, to consist of a fundamental misunderstanding of our intuitions regarding early existence. The necessities of material obligation are always viewed against our spiritual nostalgia for a fuller life, unimpeded by the burdens of vulgar practicality. In fact, by returning psychogenically to the material experiences that originally created our "otherworldly" affinities, one realizes that the world of the "spirit" is none other than the haunting recollection of our earthly innocence, once enjoyed as a poetic mystery. The mind of the child, which understands nothing yet perceives everything, is lucid without ideation. Yet, in the mind there slowly develops an intellectual need to comprehend what it sees, creating the sense of awe that it attributes to primal experience. In retrospect, the memory of this early stage of consciousness, in itself perfect, though lacking in conceptual development, explains the genesis of our otherworldly preoccupations, which are in effect the nostalgia to return to our oldest mental habits. Material responsibility follows long after the establishment of pure perception; poetry thus precedes reality, and poetry haunts our subsequent existence with its suggestions of a "higher world", forever at odds with utilitarian necessity. This opposition of dual worlds is the product of the juxtaposition of two different modes of thought: one non-rational and ancient, and the other intellectual and forcefully imposed by duty. Yet, both revolve about a common center: the earth itself, though seen from two points of view, capable once again of resolving its duality of matter and spirit into a single unity of primal perception.

Many of the problems of psychology relate directly to these perceptual mechanisms and their mode of formation. Gestalt psychology, especially, has recognized the importance of images from earlier experience in synthesizing our comprehension of reality. Investigators note, for example, that impressions formed on the retina by a series of random scribblings last longest where they resemble familiar patterns, showing that even physiological perception is governed by conceptual organization. A clue to the nature of psychotic delusions may also be discerned in these processes, which can fortunately be observed firsthand during the psychotropic descent into that part of the mind where the primal patterns are at work. The supreme role of archetypal memory in integrating our awareness of the world is again illustrated by the fact that various psychotic states are actually conditioned by the content of one's cultural background:

> Within the framework of the American culture, manic-depressive reactions tend to be evenly distributed among all socioeconomic levels. There are, however, significant differences in the incidence figures for this disease from one culture to another; for example, in

This relationship between images and insanity can actually be pursued much further than we have suggested here. It is likely that most, if not all, of our thinking is accomplished through sensory symbols that are manipulated by the imagination, since intellectual facts are more readily comprehended after they have been organized into patterns that reveal their relationship to one another. This need to simplify knowledge into comprehensive images persists in adult thinking, showing again how our earliest habits of eidetic perception continue to determine the forms of our understanding. The fact that schizophrenic thinking tends to be abstract rather than concrete indicates a failure in this ability to relate existence to these primal modes of comprehension. In the face of growing demands upon one's rational faculties, these simple pictures must be enlarged to include the present, while at the same time retaining their accustomed concreteness. Etiological factors that fix them at a primitive level, such as the emptying of the meaning of present reality through lost continuity with the authority of the past, destroy the needed means of integrating more complex experiences into similar patterns. If developing intellectual processes cannot be incorporated into these original images, which are mentally equiv alent to pure perception (i.e., uniting meaning and external form into a single sensation), the tangibility of one's objective world-picture becomes fragmentary and overly complex. For a time, one may attempt to compensate for this lost clarity with increasingly abstruse thought, but is eventually obliged to return to a more elementary stage of pre-rational comprehension. On the other hand, the inability of the mystic to adequately describe his visionary experience is explained by the fact that conceptual imagination has suddenly succeeded in reducing an intellectual abstraction to an eidetic image, through which the meaning has become apparent, and without which it again becomes a mere idea. The great value here of LSD and mescaline is that such processes are repeatable, making it possible to recapture concepts with palpable imagery, whereby their meaning can be directly intuited as pure cognition.

> New Zealand manic-depressive reactions occur two and one-half times as frequently as does the schizophrenic reaction, whereas the opposite is true of the picture in the United States. Among Kenya natives only manic reactions were observed to occur when the disease develops. The role of cultural factors in the etiology of the disease is not understood.
> (Coville, Costello, Rouke, *Abnormal Psychology*)

We can, nevertheless, begin to understand the principle of cultural conditioning upon such mental diseases if we recall an observation made earlier: Regardless of the physiological factors in psychotic behavior, the forms it assumes come from the mind of the individual himself, and these in turn, from his total experience, just as the forms of "normal" behavior come from the patterns of comprehension in the "healthy" mind. What has never previously been conditioned cannot be transformed into behavior, either "normal" or psychotic. Hence, the problem of how we think and perceive must eventually be related back to the store of conceptual experience, which can be rendered visible through means of hallucinatory representations, creating more direct methods of revealing our psychological mechanisms than those based on theory or inference alone. These applications of the hallucinogenic drug might be proliferated at will, for wherever sensory perception and understanding are involved, a means of tremendously magnifying or exaggerating the process would be potentially useful. In all events, it is most important that such researchers no longer be restricted to the mere clinical background of psychotic individuals, but extended to a general cross-section of normal society to provide a larger picture of the human mind as a universal phenomenon, through which the processes of history and culture perennially manifest themselves.

# -11-

Yet, beyond these objective uses of psychotropic drugs, there will always remain those ancient visionary gifts, for which they have been valued by innumerable races since the beginning of history. If our encounter with life can be made more splendid, even for a short space of time and at no cost to one's physical well-being, the virtue of such an experience speaks for itself. Like the sweetness of music, which contains its own meaning, the beauty of earth revealed in moments of visionary attunement require no intellectual or moral justification. Each revelation of life's potential magic strengthens us against the despair that threatens our little span of being, and each perception of a world transfigured is a higher stage in our poetic development.

The ultimate use of poetry is that we learn to see through keener eyes the things we ordinarily overlook. Its greatness lies but secondarily in its technical beauty, for it is chiefly the record of deeper experience which men of vision have encountered, and which we may also experience through poetic example. Li Po sings:

> I feel dim and wistful yearnings to be in Ch'ang An
> In autumn when the crickets sing by the parapet of the
>     golden well
> And the light frost makes one shiver and the bamboo
>     mats are cold.
> By the flame of a solitary lamp my thoughts are are like
>     to die away . . . .

And we feel more poignantly the loneliness of every autumn night. The sound of the insects and the feeling of the chill darkness are recreated as vivid patterns of future awareness. Through the eyes of the poet, the clouds, the rains, and the scent of earth all reawaken in our own souls; profounder dimensions begin to unfold behind the ordinariness of life, and with sharpened senses, we perceive the beauty which lurks on every side.

This growth of poetic acuity enables us to participate more intensely in the fortunes of our existence, re-creating that sense of religious awe that transforms the commonplace into deeper experience. Without this sacral content, things are ultimately meaningless, for, as Rilke observes,

> [Life is] ein Spielen von reinen Kraften, die keener
> beruhrt, der nicht kniet und bewundert
> (a playing of pure forces
> that no one touches, who does not kneel and marvel).

One may learn this worship through examples of primal beauty, such as the poet reveals, or the eye of the painter discerns; one may by poetic example experience sheer communion with life itself, renewing the visceral encounter with life's own mystery, which transcends all other values. The real things, the elemental nature-things become our point of contact with original memory. Through them, with sharpened vision, we once again experience the fascination of simplicity: The flowers, "faithful to what is earthly"; the dancing motes in a burst of sunlight; and the night with its draughts of pure space, all of these felt things—all of these colors, sights, and sounds—are a hymn to existence, celebrated once more within the heart's awareness.

Mescaline and LSD intensify this awareness by providing models of heightened poetic discernment. I have never forgotten the visionary splendor of those magic scenes revealed by the hallucinogenic drug, scenes more beautiful than poetry itself can suggest. The earth has been reborn, at least visually, for me; the mystic spectacle has indeed receded into the distance, where it waits to be reawakened in some possible future experience. But, the beauty of colors and textures remain everywhere apparent with a rich intensity that they never before possessed; odd bits of plastic suddenly shine with unexpected luminosity, and enamel surfaces glow again with lambent brightness like those celestial backgrounds portrayed in ancient icons. In the visible world, there has appeared a new sumptuousness: in desert soil, in the weathered surfaces of rocks, and in the pallid vapors of the sky; even the colors of a gaudy supermarket have a novel quality that reminds me that life is the ultimate Technicolor miracle.

These deeper pigments, these brighter lusters are poetic evidence of the finite become infinite. The religious awareness of earth's mysteries, felt within the blood, transforms the whole of nature into a theater of sacred forces, in which the presence of the Absolute is perceived with hushed veneration. The revolving seasons, the awesome spectacle of the sun, the storms, and the departure and return of life, all manifest a deeper sense of being, drawn from our sentient kinship with the timeless processes we have learned to behold. This is the naked religion that D. H. Lawrence imagined, the natural worship that the Greeks elevated to an expression of joy before their vision of deified matter:

> *Es sehnt der keusche Himmel sich, zu umfahn die Erdl. Sehnsucht ergreift die Erde, sich zu vermhlen ihm. Von schlummerstillen Himmel stromt des Regens Guss. Die Erdl empfangt und gebiert den Sterblichen Der Lammer Grasung und Demeter's milde Frucht; Des Waldes bluhenden Fruhling lasst die regende Brautnacht erwachen: Alles das, es komrat von mir.*
>
> (Chaste heaven longs to embrace the earth; Longing fills the earth to wed herself to him. Showers of rain fall from the quiet sky; The earth conceives and bears to mortal beings The green fields and Demeter's gentle fruits. And the nuptial rain awakens the blossoming springtime in the forests: all of this comes from me.)
>
> (*Aeschylus*, fragment 44)

Here speaks Aphrodite, the embodiment of love and desire, whom the Homeric world borrowed from the Asiatic Great Mother. From her came the moisture that sustains the fruitful earth, and the longing that drives existence to perpetuate itself throughout eternity. To comprehend the same voice in our own world is the ultimate fulfillment of consciousness because it is tangible with a meaning of its own, trembling with the immediacy of actual experience, pointing not beyond but to the mystery within the heart of Life itself.

Thus, the psychogenic vision becomes a spiritual restorative, revealing to a skeptical age the oldest religion, the one perennial remaining to us when belief in the myths of unreality is no longer tenable. Its abstract poetry is transcended by a greater poetic meaning, a meaning that speaks directly to the blood, through which we are connected indissolubly with the last atom of endless being. This reawakened awareness is the "rebirth" experienced by peoples throughout history, the very root meaning of all religion. By means of this vitalizing encounter, the past is reunited with the present; it reveals in the here and now the mystery that childlike perception attributed to natural phenomena. Mountains recapture the misty beauty of Chinese paintings; the sky recaptures its vistas of infinite space, and its clouds, their suggestion of eternal longing, which utilitarian eyes have forgotten. That supreme appreciation of simple shapes, lines, and surfaces—celebrated in Haiku poetry (Ah, how glorious the young leaves, the green leaves, glittering in the sunshine!)—is manifest once more; through visionary senses, life is finally and abundantly consummated, showing that the earth that bore us is indeed our infinite home, so hauntingly described in memories of a "higher" existence.

When Basho writes, "The fragrance of some unknown blossoming tree filled all my soul!", we begin to understand the deeper meaning of experience, for the motives that drive us onward—the needs that hunger for appeasement—reveal that man is a poetic creature, as well as an intellectual one. As long as the primal life is compromised by mere utilitarian "reality", the wellsprings of our being remain dry, and our spirits, unsatisfied. In *Faust*, Goethe wisely remarks:

> *Grau, teurer Freund, ist alle Theorie*
> *Und grun des Lebens goldner Baum.*
>
> (Grey is all theory, dear friend,
> And green life's golden tree.)
> (I, 2038-9)
>
> *Man sehnt sich nach des Lebens Bachen,*
> *Ach! nach des Lebens Quelle hin.*
> (One ever yearns for the rivers of Life,
> To reach the sources which feed our being.)
> (I, 1200-1)

Realization of primal existence, then, demands a journey into the depths of the far-off land, where amongst the primal images are the forms of everything that we perceive or desire in the present world. Returning us temporarily to the sources of our experience, LSD and mescaline enable us to recapture our dreams and to resolve that poetic paradox that consists of all the things that we know do not exist, yet whose emotional presence cannot be disputed. What the intellect denies, the heart continues to believe, for deep in our memory there still live those pictures of island paradises, those perfect settings of novels and dramas, those impossible love affairs, and those medieval romances that never transpired. Wherever imagination suggests the existence of beauty beyond the fields of evening sunlight or harmonies beyond earthly music, we stand in the presence of the far-off land. Whenever we perceive existence that we cannot touch or encounter the infinite in ineffable moments of otherness, we respond to the voice of primal life, penetrating through the weariness of banal necessity. All that is sacred and holy is derived from its ancient authority; all that is sublime reflects its distant glory. Like the unicorn, it is visible only to the virgin soul that has regained its childlike vision, for pure perception alone is capable of revealing the separate parts of existence, reunited in a single presence, in which both matter and spirit are illuminated as a timeless moment of perfect identity. Thus life again becomes concrete, its opposing forces resolved into a practical course of action, lighting the way to future encounter with the Beyond in the forms of everyday experience. For having once shown us the poetic model of our original paradise, the psychogenic vision assures us not only that the substance of our dreams still lives, awaiting fulfillment in reality, but that the far-off land is in very fact our own real world, as fresh today as it was in memory—as rich in legendary beauty as the lyrical splendor that spread itself before the eye on the morning of life's innocence.

## Notes Taken during Experiments with Psychotropic Drugs

### August 1959

10:45 AM: Took .330 grams Mescaline Sulfate

11:20: First very slight signs of visual disturbance. Slight dizziness. Nervousness

11:35: Very slight nausea, as if on board ship. Slight numbness.

12:00 PM: Extremities slightly numb. Mouth dry. Hot feeling. Tongue hard to move freely.

12:20: First brightening of colors? More nausea, but very bearable yet.

12:45: Ate dinner. Ham, potatoes, peas.

1:00: First sense of slight euphoria. Great length of time gone by.

1:05: Stroboscopic-like light fluttering. Slight nausea still apparent.

1:10: "Effects" finally starting. Room moves; center of vision narrowed. Shirt looks green from time to time. Nausea begins to subside.

1:15: Hard to move arms, fingers; hard to write or swallow. Withdrawn feeling. Things far away. Definite effects; begin to not care about nausea, which is still slight. Want to jump out of skin. Strong reflexes.

1:25: Nearly convulse at intensity of pleasurable sensations; erotic feelings.

1:27: Paraesthesias. Euphoria beginning. Hate to talk. Drunken sound? Flashing lights. Withdrawn feeling. Have seen green color in lower half of vision or with eyes closed. Saw violet lines on ceiling. Roof looked alive and pulsating like rubber sheet that distorts and stretches. This since 1:05.

1:45: Room beginning to move about slightly.

2:15: Very little change, if any. Perspective slightly askew. Things still shift about slightly. Greenish color in lower vision. Occasionally dark-luminescent spots before eyes. First slight visual effects, walked to door, legs ached. Effort to move. Forget things. Didn't know baby was asleep. Saw black spots on arms.

2:20: First slight hallucination-effects. Colors occasionally more vivid. But every symptom to date is sporadic. Body numb, delicious narcosis. No sense of temperature anymore. Effort to write brings self back to reality completely. Then lapse into withdrawn state again. Occasional green sheets of color descend over scene. (Paper has looked greenish for hour or so). Effect more definite now.

2:25: Color changes in flashes, then normal. Wood colors richer.

2:30: More intense somatic feelings. More withdrawn. Few visual symptoms. Olfactory sensations of odd smells. See little specks crawl over hand.

2:40: Patterns of ceiling flow like water (in one continuous direction). Motion more pronounced in everything.

2:50: Spoke on phone. Bright objects in kitchen look jewel-like. Saw blue button on white sink—seemed vivid and luminous. Ravenous appetite. Cold grapes from refrigerator taste divine. Body almost not here anymore. Saw purplish luminescence on chrome objects. Pencil looks very vivid. Motor uncoordination when attempting to write. Almost feel like some other existence is writing. See pencil move and feel it, as if other force writing. I sink up and down toward pencil.

3:05: Colors are brighter. My left thumb looked phosphorescent just now as I held this book (writing pad).

3:10: Psychic elation for an hour or so. But now feel for moments as if were coming out of it too soon. Mind clear. Bodily paraesthesia. Have felt as if I kept clenching muscles in body—for what purpose? Have to consciously let go tension. Pattern on ceiling flows still, if I let go long enough. My goddamn brain is too powerful to let me go rightly. Too analytical even now. As I analyze things, their visionary quality goes away.

3:15: Can easily concentrate on things. Have had no sense of mystic revelation or meaning behind things, as yet.

3:45: Height of experience. Can dream brilliantly—see visions with eyes closed.

4:10: Hear sounds as light patterns in visions.

4:40: Have been dreaming weird dreams. Sexual reality. Feel can really touch things I dream.

4:55: First feeling that effects are subsiding. Feel can grasp reality any time I want, but no wish to. Bodily lassitude too delicious.
5:15: High period of ecstasy. Feeling of immense satisfaction, joy, mystical revelation.

Till 6:15: Sheer bliss with music (Wagner). Callers broke spell. Perfect lucidity with other persons, but all seemed fresh, new, invigorating.

7:00: Wearing off. Jangled nerves to certain extent. Tendency to lapse into withdrawn feeling when I allow it. Took cup of tea and 25 mg Thorazine orally.

8:00: All psychic effects gone. Merely feel "drunk" now. No unpleasant effects as yet. Perfectly rational. City lights (in darkness) look like jewels. Very intense and colorful.

9:30: Bodily narcosis still. Mind normal.

10:30: went to bed. Tired, numb pleasant

## November 8, 1959

12:40 AM: Took 25 mcg. D-Lysergic acid diethyl amide base in lactose trituration.

1:00: Very faint sense of anxiousness, but no palpable symptoms. Believe symptoms imagined, LSD not working.

2:05: First definite sensation, certainty regarding effectiveness of dose. Feelings (which to now have been imagined) have become real. Colors seem very slightly brighter.

3:00: Height of syndrome. Feeling of physical euphoria very pronounced. Dreamy, withdrawn state, which can easily be broken. Body tense and alert, periphery of vision narrowed.

4:30: Physical feeling still maintained, although psychic effects probably waning. Poetry (T.S. Eliot, *Four Quartets*) seems to have much more intense meaning. Have been hearing music in mind, but much more palpable than usual "daydreaming".

4:50: Effects definitely subsiding, though sensations still quite tangible (Have experienced no hallucinations on small dose.)

6:00: Feel quite normal, except for slight numb sensation and a light cold sweat. Took 25 mg. Thorazine orally.

8:30: Quite normal; visitors. Occasionally, tendency to suddenly withdraw into self, feel recaptured mood of tranquil bliss. Once, flowered pattern on wallpaper seemed to be a real garden, as in a *mille-fleur* tapestry, although it was not in any way distorted. Nerves tense, but not uncomfortable.

## November 21, 1959

12:05: Took 50 mcg. Lysergic acid diethyl amide base.

12:25: Transient feelings of cramps in calves; tenseness.

12:35: Muscular weakness

12:40: Withdrawn feeling; nerves alert, muscles tense. Flushed, pleasant sensations in body.

12:50: Warm, euphoric feeling. Jaw and throat muscles tense.

12:55: Dinner

1:05: Intense euphoria; numb, drugged feeling.

1:10: First brightening of lights in periphery of vision. Intense colors. Erotic sensations.

1:20: Listening to French popular singer on tape. She seems to be in room, breathing over my shoulder.

1:45: Intensification of significance. Can look at picture and lend it great depth of reality. Otherwise, life seems suspended in space and time, very neutral, calm. (Almost zero grade of "Affect").

2:00: In-womb feeling of warmth, calm, pleasantness. Vision narrowed, delicious narcosis. Occasional flickering of light in periphery of vision. Will numbed.

2:07: Somatic symptoms have increased during last half hour. Colors occasionally appear brighter in flashes, but overall appearance normal. Tendency to withdraw. Attention periodically arrested by conspicuous or bright object.

2:30: Feelings of elation; great plans to do things (in future); expansive mood. Things seem novel and new; they create desire to smile or laugh. But overall sense of narcosis and benumbed will. When I sit down, I immediately have desire to withdraw into warm, dreamy state.

2:45 to 3:00: Conscious repression of psychological material seems to be relaxed; wish to express inner feelings (communicate), but presence of others seems to be set up resistance. Flashes of light; wallpaper flowers (roses) look lovelier. Psychological insights gained: seems that anxieties center on inability to realize imagined or promised fulfillment of experiences in life. (Things cannot be what they would seem.) LSD gives palpable reality and power of imagination to thoughts, thus bringing longed-for realization and fulfillment.

In normal life, must break down feeling that things are necessarily let down, and try to realize their inner significance (cause them to yield themselves on face value, without resisting them by anxiety that they will be disappointing)

3:15: Ability to daydream very heightened. Colors all have aura about them, but otherwise not exceptional. Mescaline visions not quite realized. Feel more relaxed than with mescaline, but still rather tense.

4:00: Physical symptoms still pronounced. Feel that psychic peak has been passed (this turned out to not be the case)

4:00 to 5:00: Period of extreme bliss and peacefulness while listening to music (Vaughn-Williams). Recapture of feelings from childhood while gazing out of window. Colors in landscape seemed rich and harmonious. Everything tranquil. Sense of infinity and eternity, of finality and ultimate peace. (The latter seemed to be connected with a sense of life's awesomeness, a sort of "tragic" awareness, yet not frightening, but touched with the beauty of death-wishes.)

5:00: Physical symptoms still maintained. Took 200 mg. nicotinamide orally, to test effect. If any. Still tendency to withdraw, when allowed.

5:30: Went for walk in dusk. Winter scene had novel freshness about it. Seemed to be pervaded by a long-forgotten childhood sensation. Lights seemed garishly bright.

5:45: LSD effects only slightly, if any, suppressed by vitamin.

5:55: More definite clearing of mind. Difficulty in lapsing into withdrawn state, although still pronounced. Sense of calm and relative disinterest. Vegetative symptoms seem to be disappearing. Cool feelings, in contrast to previous warmth.

6:45: Effects completely past. Feeling of happiness and ease. (Nerves slightly jittery until bedtime, but not unpleasantly so. Took 100 mg. Pentobarbital to sleep.)

## December 17, 1959

11:45: Took 75 mcg. Lysergic acid diethyl amide.
12:10: Ate lunch.

12:20: First physical symptoms; slight dizziness, weakness, mild anxiety, narrowing of vision.
12:30: Definite signs of drug action. Ability to withdraw (into Self) already beginning. Weakness of fingers while writing.

12:45: Euphoria becoming pronounced.

1:05: Effects intense euphoria to consider in full course. Colors seem more intense, bright, upon changing scene. (One adjusts gradually to surroundings so as not to notice extra value of sensory perceptions. Sudden change, however, is filled with novelty.)

1:15 to 1:40: Erotic sensations. Intense euphoria.

1:40: Took additional 25 mcg. Lysergic acid diethyl amide. Perceptual changes very slight. Occasional pulse in wallpaper. Mostly withdrawn state. Colors sometimes have halo. Things in periphery of vision seem at times phosphorescent. Time seems to slightly lose meaning. Am aware of time, yet it seems suspended and meaningless.

2:20: Very intense symptoms, euphoria. Still very little perceptual changes, though all has a new beauty. White things have a violet aura. Had "electric feelings" (noises and muscular twitchings . . . all statically charged); slight parenthesis. Bizarre imagination. Can "daydream" odd phantasmagoria. Ability to withdraw marked. Slight brown-orange glow about fingers as I write. ** Produced first hallucinations as burst of colored crystals, which subsided into a whirling band of color, gradually coming to rest (Synaesthesia).

2:30: First kaleidoscopic color visions with eyes closed. Brilliant, fantastic daydreams. Molten colors, lights, geometric forms. People play a large part in visions. Brilliant pantomime gestures; flow of color and matter. Still easy to return to normal if desire. No anxiety whatsoever. Body tense.

To 3:15: Extraordinary vision of life, all pulsing with fire—Heraclitean ultimate reality. Was one with cosmos; felt inner being radiant with heat of a million suns. Cosmic dance. Became personified in human forms. Persons then gradually became younger, as I regressed back through womb, became one with eternal slime and sexual generation. Was surrounded with swimming sperm-like animals. Felt self rising through earth into new plants. Was one with all life and Being? Was God, in God; slime of eternal fire bursting through everything. Pulse of existence shattered into billion sparks of divine creative ecstasy. Indescribable religious revelation. Am color, breath, substance of million warm faces in vision. Some other hand writes. I merely look on from Parnassus height. Oh! Beyond description! If other foolish swine could know! Tongues of flame impressed upon every pulsating, crystalline surface of Zenhaiku reality. Poem! Warm sex-beauty, loving contours of life in every solid substance, yielding themselves blissfully to each wave of undulating creativity. 3:40: Anxiety that children will soon return. Am at height of ecstasy. Wonder—should I take antidote to see how it works at such a peak? Fingernails look daubed with red paint. Dynamic quality of vision imparted to everything in room. This could go deeper. If I were L**, I'd be frightened. But I'm myself, and my metaphysical enlightenment leads me to embrace this chaotic vision of everything being drawn into the vortex of fire and flame. This is ultimate reality. God!? I am God!

3:45: For scientific reasons, I will now take antidote, although I don't personally require one. Can hear crickets chirping—am back in childhood scene—Summer evening—this most unusual! I'm just getting hang of really projecting myself into past. Can dream anything, be anywhere. Creative ecstasy. Brilliant fire bursting in vision. All throbbing in tune with inner pulse . . . which is one with All. Electric nerves flow from self out into cosmos. Can feel pulse of Eternity running through body. Drug has removed barrier of ego—Reality comes through, or is rather freed from outer restraint and pollution, so as to be clear, transparent and vivid with light and fire. Hand is writing furiously—feel power to leap over any impediment. (Note added later: Handwriting at this point was actually more decisive and well formed than normally.) Fewer to project myself in Divine creative act. Have never before been whipped to such a frenzy of ecstatic God-like joy and positivity. Everything is outgoing affirmation of light and generation. Thoughts pouring out like molten lava. Body is luminescent mass of primal substance. Oh the joy of it!

3:55: Notice have neglected to toke antidote. Hand is hollow cave from whence issues green luminescence. Am hungry. Tremendous physical will. See all instincts such as hunger personified as cosmic necessities, all pert of Life. Physical symptoms: difficulty speaking, mouth gelatinous and sticky. Perceptions as if drunk, but mind very clear. Only have to give pencil a nudge with brain and it writes by itself. (Note added later: I recall only moving pencil across paper, "wiggling" it up and down. The letters formed themselves automatically in obedience to my intention, but quite without conscious effort of a motor sort.) Great intellectual strength (imagined?). Perhaps ability to function normally will be impaired if I try.

4:00: Taking 300 mg. niaclnamide to objectively test situation. All colors in other room burst in vivid splendor. This is maximum degree of psychic consciousness—change that I have ever experienced. It is hard to write (as a conscious effort): hand looks odd doing it for me. Mind clear but flesh weak.
4:15: Talking to wife maintains perfect contact with reality. If I pause for a moment, I instantly relapse on my very feet into withdrawn, visionary state. Had glass of punch. Red liquid in plastic glass, with deep blue flowers on sides: Color so divine it almost makes one weep! Never saw colors so vivid and eidetic, even under mescaline. Face flushed; sweaty skin. Eyes dilated.
To 5:15: Listened to Mahler's third symphony, last movement. Vision of absolute sublimity, peace, splendor. Colors of wallpaper seemed indescribably rich and mellow; ceiling of mother-of-pearl and silver. Walls of silver and maroon flowers. Like lying in a magic cavern, lower end of which opens onto mysterious and beautiful casements. All pervaded with sense of rich, peaceful harmony. Little trickles of

water flow between flowers on walls; all breathe and pulse. Every line of music corresponds to inner emotional flux. Mere physical reality of a rectangular room is a splendid reality beyond comprehension.

5:20: Little effect, if any, from niacin. Still no anxiety, could go on for eternity. Perhaps slight lessening of intensity of state. This first sign of decline. Physical effects as strong, but vivid period definitely past. Mental effects losing luster and novelty. Mood corresponds to sunset grey, but no apprehension, only approaching zero-grade affections. Tendency to withdraw still pronounced.
Wallpaper flowers ripple and undulate on waves of water. Muscular tension has become rather tiresome; feeling of body-fatigue.

5:35: Took 50 mg. Thorazine orally. Don't believe niacine enough to completely alleviate symptoms, although it has definitely (!) contributed much to restoring normal contact with reality (LSD syndrome has lost its dynamic effects, but retains passive physical symptoms of lessened perceptual acuity). Colors have lost inner glow. Greater ease in establishing contact with outer world.

5:45: Psychic disturbances virtually completely gone. Only muscular tension and tiredness remains. (Thorazine has not yet had time to act. The above state due to niacine).

6:15: Normal state returning fast. Nerves beginning to relax under Thorazine. Mental state clear. Overall feeling of drunkenness, except for thought clarity. Ate dinner.

To 7:15: Listening to Strauss's *Arabella* by Christmas-tree lights. Branches still move slowly. Recreated palpably the sensation of Christmas as knew it in childhood—all this so real as to be unbelievable. Have not known this for twenty-five years. Whole epoch of Old Vienna was real and present. (Note added later: This corresponded to a fancy experienced in actual childhood at about eleven years of age.) Childhood fantasies relived.

7:45: Perfect normality at public affair.

## January 3, 1960

11:10: Took 100 mcg. Lysergic acid diethyl amide.

11:45: First physical effects beginning.

12:00: Feeling of drunkenness. No euphoria as of yet. Slight visual impairment. Colors beginning to be intense; stroboscopic light effects. Rather rapid onset of symptoms. Very slight malaise.

12:05: Flashes of light in periphery of vision. Malaise passing.

12:10: Euphoria setting in. Erotic feelings. Perception altered. Muscular tension, slight movement in field of vision. Withdrawn state beginning. Depth and dimension altered.

12:20: Ate lunch. Colors very distracting on plate.

12:40: All colors very bright as in Kodachrome pictures. Very intense feelings (of symptoms).

1:00: Ceiling patterns flowing like water, stretching back and forth. Listening to Mozart music. Indescribable bliss and ecstasy. Colors of walls exquisite. 1:30: Colors seem slightly less intense. Things suddenly lose interest and novelty . . . but begin again when I lapse into self.

1:45 *****

2:00: Physical action precipitated period of great sadness, emptiness. No meaning to anything. Thought ability to dream gone. The warmth and outgoing positivity of previous experiences lacking now.

2:30: Can still dream vividly. External colors have lost some of their splendor. Somehow, whole experience, though at times very pleasant, dominated with utter meaninglessness of everything. Purposelessness of existence personified in feelings. (Added note: This part of experience evidently coincided with schizophrenic lack of affections, although nearly perfect lucidity had replaced the hallucinatory phase just preceding it.)

2:40: World seems cold, strange, empty. Can't make inner self work this time. No anxiety, just utter zero-ness, raised to almost mystical intensity.

2:55: Had a period of warm feelings, which have again passed. Plain simple fact is that rationality has prevailed too strongly. Effects are wearing off. Too many lucid periods. Dynamic qualities of hour and a half ago definitely gone. World is just what it is—a cold, winter afternoon with glaring sunshine meaning nothing. Inability to keep up illusions. Not LSD that's wrong—it's the damn world outside and the horrid reality to which you have to keep returning. If I've lost this means of some pleasure in life, what's left? Still feel drunk, but mind inside doesn't work anymore. Visions gone. Fire gone. (Added note: This seems in retrospect to have been an expression of general repugnance for cold reality, and the knowledge that the beautiful illusions of past beauty are fled forever. Man cut adrift from his erstwhile fountains of mystery and dreams.)

3:10: Effects seem definitely subsiding by themselves. Feeling of general disappointment. Perhaps mood of morning (had slight hangover from 50 mg. Phenobarbital taken late previous night) predisposed me to experience without inner warmth. Worries regarding university and beginning of quarter. Fag-end-of-holiday spirit. All pleasures over, misery beginning soon. 3:30: More blissful experiences again. Beautiful kaleidoscopic visions, electric—fire feelings. All splendid in inner mind.

3:50: Splendid dreams—all beautiful and exciting again. Sound outside burst into million round bubbles of color—gold against green. Field of vision with eyes closed is lovely purple velvet, pulsating with brocade patterns. World outside seems illuminated with gigantic sunlight-force, which goes through everything like crystal, making it radiant as if molten glass, resplendent with rainbow-sparkle tongues of fire. Euphoria returned, along with renewed ability to withdraw.

4:00: Thank God ability to experience has returned! Life has recaptured meaning. Feelings of happiness. Period of inactivity seems to have brought back inner fulfillment in imagination, 1:45-3:15 period of physical activity, moving around, etc. This against background of enforced reality emptied all of its potential ego-fulfillment. Definite proof that in order to enjoy LSD or mescaline, one must have opportunity to withdraw from outer distraction. Outer world is enemy of self—not the ego. On closing eyes, all immediately flames up into visionary fire and splendor. Objective sight has, however, lost the hallucinatory quality it had at 1:00 PM. Ability to re-establish contact with reality complete. Has perhaps been more so throughout whole experience.

4:20: On looking out front window, world looks indescribably fresh and vivid. Feel as if I have returned by painful effort from Hell of absolute meaninglessness into reassertion of divine beauty in all. Worked up from zero-feelings, through feelings of last fifty minutes, and now see world outside resplendent and meaningful.

4:40: Took 25 mg. Thorazine orally to complete easy transition to reality.
5:00: Lay on couch for last fifteen minutes, looking at scene in front room. All unearthly-beautiful. Indescribable peace and harmony, lovely richness of textures. Light and color serene, deep, Haiku-tangible. Mere flat wall surfaces so meaningful and beautiful that no words can express it—only to look at it and know! Tranquil feelings; peace, oneness, mystic bliss descended into earth and stone to reveal its transcendent Dharma-body. (Added note: An obvious reference to Aldous Huxley's description.) The mere ordinariness of it is the ultimate revelation—because the ordinary becomes (or reveals itself as) the Real, the Beautiful. Had cup of tea.

5:20: Feel renewed flood of enthusiasm, euphoria, bliss, joy! Caffeine seems (?) to act like Dexedrine under tail-end lagging effects of LSD. No hallucination, but indescribable inner warmth and desire to communicate. Very strange after the middle period of depression (rather, lack of feeling!). This perhaps a new dimension of experience; almost the best of it! Pencil now seems to have life of its own, move almost automatically. Also, certain tendency to withdraw again, almost as if whole experience were beginning again, with new beauty and inner radiance, except minus visual hallucinations. Oh! The beauty of it! The joy of it! The meaning of real harmony (here seen and felt, tangibly, not as an abstract idea, but as reality!). Peace, contentment, serenity. Hesse's *Heiterkeit* revealed to me in waves of bliss.

(Immediately afterward, went for walk; felt like youth again in old childhood neighborhood on a long-forgotten winter afternoon. Mountains reminded me of those seen hovering like pale blue mist in a Chinese painting. Symptoms of syndrome appear to have been actually subsiding at 6:00 and to have been completely past by 7:00.)

## January 23, 1960

11:05: Took 380 mg. Mescaline sulfate in half glass water.

11:20: First signs of effect. Nausea beginning.

11:45: Very marked nausea. Had to lie down to suppress feeling. Great depression and feeling of regret at having taken drug.

12:00: First sign of brightening emotions. Regret passed.

12:10: Can dream vividly with eyes shut. No hallucinations, as yet. Colors in room clear, serene and pure. Harmonious feeling. Nausea subsiding. Erotic dreams.

12:45: Begin to experience slight withdrawal. Colors bright, fresh, vivid. Hunger sensations. Nausea almost completely gone.
12:50: Ate lunch. Hyperreflexia, muscular tension, drunken feeling, almost catatonic state of withdrawal when I retire.

1:00: Very drunk feeling, but sense of reality very little altered. No visual hallucinations, except for reflected extension of colors across lower field of vision. Feeling of warmth and flush, perspiration.
1:15: Gorgeous colors in vision; sparks of light in periphery of sight. Euphoria beginning. Positive phase of experience definitely setting in.

1:20: Sheer bliss and ecstasy; very intense euphoria. Brilliant colors in view with eyes closed.

1:30: All splendid fire and heat; divine force pulsating in vision beyond all description. Gold, red, orange flames blending into harmonious patterns of incandescent beauty.

1:50: Absolutely indescribable visions; all in most utter, violent motion, throbbing, pulsing with electric energy. Things changing from one scene to another without cease. All in bright, iridescent colors. Fluttering lights, firework displays; body tense, "galvanized" with electricity. Hallucinations splendid. Feel can will anything into reality. Hand while writing looks detached; imagine I can merely write a word down and the idea will be real.

2:15: Vivid dreams; imagined could turn head inside out, see backward, etc. Very tense. Feel creative energy, revitalization. Colors outside very vivid and beautiful.

3:15: Period of indescribable recall: actually relived childhood in every detail. In successive scenes, seemed to be growing younger, until age of approximately seven or eight. Could smell the little nook north of house with its bushes and greenery, the bark up in boxelder tree, taste the apples on roof of Mr. Moore's shed, smell clothing stored in upper hall closet, see shady lane down avenue, trees in yards where I played as a child (complete with every knot-hole and configuration of the bark); remembered imagined games played with old cat; remembered suddenly dusty ledge outside bathroom window, and feeling of climbing up to it while creeping around house to my own window, dusty screens outside of room, etc., all absolute and real—never any memory remotely like this before! Felt bliss of never-ending summer afternoons, warm, damp greenery under trees, looking up at clouds and bowers of leaves; lying on warm cement under eaves on summer afternoon and scent of rain; no anxiety, ever-present childhood eternity of sheer poetry in timeless, lazy afternoons of July . . . Oh! The dear earth and scent of earth that bore me! Was back in crib in mother's bedroom, watching blue-emerald light of summer afternoon outside in ancient landscape . . . bathed with bliss of mother's smile . . . time hath no meaning for us—eternity is ever and now, and happiness is easy, not strived for. Oh—the innocence of a little boy, without cares, worries! To know again after so long what it feels like! Thank God for a bounteous blessing like mescaline to recall to mind what is more important than gold and hurry! Something I thought forever lost to me. The odor of rain on dust, of sap in heavy, lazy, dreamy boughs, the scent of earth through blades of grass, seen by child with face pressed to the ground—caressed by endless, rolling-green, English-father's-house lawns! Sun warmth and fertility, oneness and peaceful, desireless harmony. God! I can recapture anything, just for the asking! What a miracle! Oh, joy and bliss sublime! Everything crowds upon me like an immense embrace, loving, selfless, innocent!

3:50: Muscular tiredness, tenseness; deep feeling of mescaline intoxication. No objective hallucinations this trip—walls very steady, etc. No feeling of loss of reality. Split personality sensation marked—very clear mind, but able to dream and experience visions at will. Little heightening of colors, as with LSD, much less visual hallucination, although inner vision even more fantastically alive.

Mystical sense not as noticeable; more an objective phantasmagoria. Continuation of warm, inner elation and electrified senses more noticeable than in LSD syndrome. Would conclude that mescaline is more of a nervous stimulation, and LSD, more of an actual psychic change. Withdrawal of mescaline more catatonic than LSD; tension much greater. The chemical relationship of mescaline and the amphetamines obvious.

4:15: Brightness of hallucinations (inner vision) subsiding. High point of visual acuity passing, but not gone. Colors in vision are much paler, lacking in reds, greens, etc. Now mostly blues, greys, transparencies.

4:40: Still able to dream fantastically, but in subdued colors.

5:00: Took 50 mg. Thorazine.

5:00 to 5:30: Listening to music; one understands Mozart for the first time—as each phrase unfolds, waves of contentment—so pure, so serene—pour their lambent light upon the languishing soul, drunken with sheer bliss! Every silvery sigh, every shudder of peace and fulfillment rises clear and transparent before the senses, like bubbling water in mother-of-pearl chambers! The pure, oily, gut-richness of string tones sound with poetic sensuousness that they never before possessed. Every pulse of an instrument is a perfect sigh, every nuance an expression of perfect emotion. How intricate and lovely the interweaving of naked sound—the pure presence of eternal harmony and experienced sensation! The incomparable splendor of Mozart revealed as never before!

5:50: Effects beginning to subside. Cup of tea produced cozy, warm feeling. Muscular tension, slight cramps all over body; drugged feeling. No unpleasant sensations as yet.

6:00: Feel elated; rich joy of surroundings, family, of many possessions. Harmonious emotion of outpouring good will and happiness. Feel have been through an uplifting experience

### February 28, 1960

2:30 PM: Took 25 mcg Lysergic acid diethyl amide to test more exactly its qualities as a short-run euphoriant and mood-transforming drug. This followed a dose of 15 mg. dl-desoxyephedrine, taken approximately at 2:45, the effects of which might well alter those of the LSD. The purpose of this experiment was to verify the possibility of directing small doses of LSD into short, non-anxiety states by a previously administered mood-ameliorating chemical.

3:30: A familiar object in the room unexpectedly took on erotic undertones. This was all the more surprising, since no other noticeable effects had as yet occurred. It also seemed that colors, though no brighter, were subjectively more striking, although this may have been due to suggestion.

The first definite symptoms of the drug were vegetative; at about 4:00, tenseness and sweating, with slight euphoria, began. By 5:00, my mood became markedly altered, as I withdrew into a pensive, dreamy state. A strange world of otherness, neither euphoric nor dysphoric, ensued. It seemed that some unknown mode of reality lay thinly hidden behind a veil, whose fabric I was not to penetrate. An unearthly tranquility pervaded the scene outside: winter afternoon, sunset; all splendid in the orange light of the waning day. Brick-reds of houses, the rich grey of the cleared streets; pinks on the snow, beneath a cool, pure-blue sky. Faint wisps of clouds in the far west, all rose and mauve colored, suggestive of some far-off, mysterious land. Everything etched in sharp, jewel-like clarity, inexpressibly beautiful, calm and harmonious, as in a vision of ultimate, changeless Unity. The sense of unknown meaning, distracted dreamlike withdrawal, and nameless longing, almost devoid of emotion or feeling, made the illusion of impending mystery all the more remarkable, as if one were standing at some unthinkable Antipodes of human imagination, beyond time and space, yet convinced of the harmony and beauty of the experience. All of this recalled a childlike state of complete humility, wonderment, and incomprehension.

Listening to "Tristan" excerpts, there suddenly arose in my imagination the awareness of childhood recollections, trying to draw me back against my will into their sinister magic. Memories of my old home unexpectedly materialized and presented themselves as some sort of death-wishes, viewed as a flood of overpowering, negative urges. The music was powerfully suggestive, at the same time, of tragic-sublime visions, in which death, beauty, tranquillity and world-mystery were inseparably united.

6:00: Took 25 mg. Thorazine. My state of detachment and deep reflection lasted until approximately 6:45. No impairment of reason or perception noted at any time, slight muscular tiredness remained afterward.

## March 5, 1960

11:25: Took 75 mcg. Lysergic acid diethyl amide

11:45: Slight withdrawal beginning, rather definite euphoria noticeable.

12:00: Very intense euphoria.

12:05: Rapid onset of symptoms. Slightly intoxicated feeling. Narrowed vision.

12:20: Waves of elation and pure bliss. Euphoria has grown steadily since taking drug. Little change in perception.

12:30: Euphoria almost overwhelming. Great tension and elation. Physical sensation one of ecstasy.

12:40: Hyperreflexia, even greater pleasure. First notice that lights are slightly brighter. Ate dinner.

1:00: Took additional 25 mcg. LSD. Feel peak of elation, euphoria. Colors beginning to play their fantastic illusory images around periphery of vision when I quickly cast eyes across field. Stroboscopic patterns on occasion. Listening to music, note that contrapuntal lines are more acute. Intricacies of music in clearer detail than ever able to notice before, as if "hi-fi" were in even greater fidelity. Warm, rich bliss. I have needed reassurance of this sort, for some time, that pleasure and happiness is indeed possible. Pencil begins to act by itself. As yet, little alteration of consciousness, only heightened enjoyment of present mode of reality.

1:35: Withdrawal tendencies alternate with perfect normality. Colors are now rich, deep, and lovely. The textures of surfaces seem richer than ever before. A face on a magazine seemed real, changing, variously smiling.

1:40: First signs of hallucinated dreams and visions with eyes closed. Daylight seems splendid and pulsating. First physical euphoria going over into psychic effects. All colors in room are vivid, deep, and full of subjective richness, especially wood-grains, carpets, and furniture. Mood would be pensive, were it not for elation and nervous stimulation.

2:00: Confusion and bustle in house keeps me constantly in touch with reality, dispels tendency to savor experience. Could be seeing visions now, if were allowed. Note perspiration for first time. Waiting for caller to arrive and leave—due in half-hour. Intolerable urge to have visit done with so as to enjoy experience.

2:20: According to previous accounts, I should now be at the peak of experience, which would be so, were it not for waiting out this damned interruption! Mood to now intensely euphoric, all bright, pleasant; here I sit, waiting for opportunity to enjoy it! Have been up and about throughout.

2:30: Sat down in retirement for a while—ability to dream is pronounced. Kaleidoscopic fancies very subdued, due to anxiety of expected visit.

2:50: Can now have a while to myself!

3:00: Too late! Height of syndrome passed while unable to enjoy it. No color heightening, synesthesia, visions, etc. Euphoria mild now, pleasant, but has no particular novelty.

to 3:45: Thank God! There is something left in nervous system! I lay alone on couch; I had splendid dreams and scintillations (though not as striking as they would have been earlier). Felt I was solid rock lying on bed, bursting with atomic energy. Matter-and-energy-image all splendid and luminescent. Saw self hurled by atomic explosions out into cosmos, past uncountable galaxies of light and beauty. Geometric patterns of color fluttered past in unbelievable profusion and delicacy. Feel very withdrawn, intoxicated.

4:00: Light! Pure light! Just to sit and stare into translucent Being! New interruption on phone.

4:10: Sitting in kitchen, gathering my thoughts. Colors are vivid and fresh. Feel equilibrium between self and reality, with general acceptance of objective world on its own terms. Have again felt childhood lurking beyond the horizon, enticing me back, but with lessened persuasiveness. Beauty in life has indeed for me been equated with childhood; reality on this spring day is selfsufficient, must be accepted in order to go on. Disenchantment has no terrors. Somehow, whole of day's experience has been subconsciously pervaded with Kerouac's novel, which I just read (*On the Road*). It seemed in visions I was reliving the frantic, masculine search for adventure and meaning; that meaning was to have been soft, feminine, a realization of all life's longings, but it was hard, bony and masculine, therefore restless, tortured, unfulfilled! That is essence of male's being! We long to be hermaphroditically united with feminine, mother, woman, sweetheart, for softness to complete our masculine torment, to be caught up in some

higher uniting, which I see in vision as life, childhood, adulthood, matter, energy, womanness and maleness, on cosmic scale

4:25: The enemy of life is the intellect, the cold, analytical intellect! I can feel it gnawing away at beauty and turning it to stone, like an acid spot eating out through a peacock and mother-of-pearl fabric. We have no ability to believe; we only know, and in this life, that is not enough. All around me in the warm, moist air are recollections of the past, which my monster-brain repels. Oh, this is all a farce, trying to pretend that anything extraordinary is happening. I can't see how other persons could get undone by LSD. I can't go far enough!

4:30: Pleasant, harmonious feeling of being suspended in time, space and meaning. Neutrality is essence of the reunified halves of existence! But living, pulsing neutrality, fire from within, divine force in perfect harmony.

4:45: Oh bliss and ecstasy of sheer reality, family, possession! Colors of cloth (children's new coats; dazzling and jewel-like. A mood of elevated, calm tranquillity. Experience this time has been throughout on level of sheer ordinariness, therefore real and divine, not one of fantasy and otherworld hauntings. Merely sit and survey all with rejoicing, noting ordinary shapes and sounds and colors, all good and tangible and rich. This has been an uplifting testament of reality; the

LSD intoxication, by being tangled with dross, eventually elevates the ordinary and affirms it.

4:50: Took 50 mg. Thorazine.

5:15: Oh warm joy and tranquil happiness! Having cup of tea, listening to *Parsifal Good Friday Music*. The sheer, outpouring affirmation of my mood and the music indescribable! Colors all bright, serene, the music peaceful; body warm, relaxing and euphoric.

5:45 Lying here in my mother-of-pearl chamber, listening to Strauss's *Vier letzte Lieder*. The beauty of this experience makes these stupid words useless and superfluous. How beautiful music can be, and how wonderfully the emotions can be molded by it; this fact beyond description. Every note drips honey-sweet and rich-like nectar, so palpable, so tangible!

6:15: Listening to Strauss's *Also sprach Zarathustra*. Saw world as an immense Hietzschean mass of life, solemnly affirming existence. Ceiling dissolved in mass of grey clouds, angrily swirling in birth and death-struggles of Being, majestically yea-saying; *I am life! I am "the force of eternity, o God!* What divine revelation of that which is! The entire cosmic drama unfolds before ray eyes in waves of Eternal Recurrence! During that drunken "Dance-Song," my soul followed every line of the music with frenetic urgency, writing, contorting, intoxicated to an almost unbearable degree. Brilliant mental images formed and passed in mad, Dionysian motion, subjectively vivid, yet without sensuous impression. As the music tolled that awful twelve-count of eternity at its climax, I experienced a sort of cosmic, spiritual orgasm, quite devoid of erotic content, yet in a very real, spiritual sense. The whole of the universe was exploding in dynamic frenzy, and ray emotions discharged madly along with the sounds. As the "Night-Song" descended, an image of Rilke-like immensity of space and distance remained, with some distant light shed over the luminous darkness. My soul hath contorted in agony, and reached a climax with Eternity!
6:45: Calmed nerves, effects definitely subsided. Still pensive and withdrawn.

## October 22, 1960

2:10 PM: Took 50 mcg. LSD (Had previously (9 AM) taken 10 mg. dl-desoxyephedrine).

3:05: Euphoric sensation in calves of legs. Suddenly noticed a beauty in the sky that I had long forgotten. A cloud formation recalled some distant notion of adventure. No color changes.

3:15: Increasing euphoria; slight color brightening; withdrawal feeling beginning; slight narrowing of vision.

3:20: Sharpened senses; all stimuli magnified somewhat. Imagination seems very strong and vivid.

3:25: I feel as if my soul were coming alive again for the first time in long, grey months. Mind working as it used to, years ago! All perceptions clear, unperturbed, bright and satisfying. Some inner, secret part of mind is active once more. All miseries and boredom of past year seem to be melting away.

3:40: Hyperreflexia in calves and abdomen. Transient tremors in hands. Great lumbar-thoracic stimulation. Mind calm, withdrawn.

4:00 to 4:15: Lay back and relaxed, with eyes closed. At first noted nothing in visual field; gradually, purely mental thoughts began to assume tangible shapes and forms; soon, was "daydreaming" very vividly, with amazing series of plastic images melting one into the other. All colors very delicate and luminous, yet not so eidetically acute as to render them physically palpable. Mostly transparent fantasies of the mind, which remained at a thought level. Almost unbelievably productive, yet illusive. Saw gleaming white thighs of beautiful women; descended into slime of visceral, internal cavities, all red and wet, warm and palpitating, like child in womb. People come and go like pale, shimmering ghosts.

4:30: Very dreamy, withdrawn feeling. Slight visual distortions, which are novel and interesting.

4:40: Took 300 mg. niacinamide.

4:45: Went out doors—suddenly emerging into a world of unbelievable splendor and beauty. Colors so dazzling, fresh, and meaningful, as if had entered into a new dimension of existence. A bed of flowers so breathtakingly lovely that words do not describe them: brilliant jewels on dark green velvet. Walked to intersection; hustle and bustle of American life suddenly assumed visionary proportions: violent motion on all sides, terrible noise. Still, all colors have new values. I see everywhere new and breathtaking combinations of subtle beauty and delicacy.

5:00: Sitting atop a rolling field of grass, like livid emeralds at my feet. The great bowl of hills beneath blue sky is etched in supreme clarity, palpable with the warm tones of brown and purple, serene and calm beyond description. Scene recalls day many years ago in South: a warm, spring evening, the fresh scent of earth rising in the air, all pervaded with exquisite longing and poetry. Every color, perfume, and touch of breeze is animated with new life, perceptiveness, the return of spring in autumn, in no sense elegiac, but happy, affirmative, and altogether unexpected. How wonderful to be sitting here once again, watching the sun go down, knowing that earth had a pain made itself conscious in which I am benign toward earth like a careless bird! Sun and soil are rejoicing in my awareness!

5:25: The moment of sun's departure! Like a religious act. The mystery of Being spreading through the sky like a great drama. At the instant of last illumination, all colors suddenly changed: transparent whites and browns, ghostlike beneath my feet. In the far west, arising in vaporous light, there shimmer! The world of my childhood longing . . . that forgotten land of Celtic and Oriental dreaming, all peach-misted and suspended in the realm of Light.

To 6:15: I have been walking through the Garden of Eden these last hours . . . or have they been minutes? I left the high lawn and began to walk home. Gazing upon a field of uttermost enchantment, I suddenly realized that this was an erstwhile rubbish heap, now transformed into exquisite surfaces and textures, rich in their own essential being! Behind me lay the mountains, still pink at the top with jewel-like sunlight. My attention was so arrested that I had to turn back and walk toward them, knowing that a veil had been torn from my sight and that I was seeing them in primal, perfect clarity! There was no distortion or intoxication about this; only the knowledge of clear vision, undisturbed and clean, overwhelmed by the beauty of the sight seen in stark reality for the first time. I turned down a side street and began to wander where my fancy took me—enchanted on every side by unexpected glimpses of paradise. Here, a violet rooftop; there, a blaze of luminous flowers; there, a mass of green against a glowing evening sky. The sense of life pulsing everywhere was satisfying and organically stimulating. Human happiness seemed to express itself as I had not perceived it since childhood. At the bottom of the road, the older trees pressed in like a primordial forest of vivid gold, sammet, and jade, magically moved by lukewarm draughts of air. The green and blazing autumn fluorescence had now assumed another aspect, heightened by the novelty of recreated childlike wonderment, which I absolutely and fully possessed. Overhead and on each side was a virtual orgy of sensuous impressions: streetlamps seen in the dusk as solitary, luminescent gems in the trees, and mysterious objects glowing in thousands of darkened recesses, under a somber-burning sky. At one moment, a manor house appeared out of some old English novel, and then, a Southern mansion, with white columns and portico gleaming in pure alabaster. Every garden, every grove of trees seemed to be a corner of a re-created Eden-world, so palpably alive that my sight could only riot drunkenly from one unbelievable scene to another. My body was light? and my movements, effortless, as I was drawn down street after street, unable to resist each charming new beauty which unfolded before me. A garden of castor-vines became a tropical forest, hung with phosphorescent jewels: Every scent and tactile sensation was powerfully magnified, in consequence of the vision I possessed, so that nothing was withheld from this unimaginable feast of the senses. I felt immense pity for the few solitary persons who passed me, who could never know what beauties I beheld. How magnificent to have the Veil of Pan torn from my eyes; to see unspeakable loveliness in every color, surface, shape, and texture, all drunkenly outdoing each other in pristine intensity; and to be wandering as a child enchanted, seeing the very heart of Being with pure, unadorned delight, intoxicated with sheer pleasure! This drug must be reclassified as a soul-energizer!

6:30: Home, in living room. The deep red drapes are alive with fire and motion. German folksongs on phonograph overwhelm me with emotion, as if I could scarce endure them. Have no trouble communicating, but withdrawal and somatic feelings still deep. Taking 50 mg. Thorazine and tea, the latter producing a rich, warm sensation of joy and contentment. Leaving for town in car; concentration seems adequate.

Later: Ate more than usual; sensations rapidly abated. No muscular tension this time, as experienced almost universally before.

## Commentary

1) The following brief description is added to satisfy the reader's curiosity regarding the more clinical aspects of the hallucinogenic experience.

Mescaline, as found in the peyote cactus, is a colorless oil, which rapidly absorbs carbon dioxide from the air to form a crystalline carbonate. It is synthetically produced as the sulfate salt, which has a bitter, pungent taste and a limited solubility in water. Pure LSD is a brownish powder, melting just below the boiling point of water to form an oil, which can be readily absorbed through the skin. Commercially, it is available as a white tartrate salt, freely soluble in water, in which form it is decomposed within a few days by the presence of air. Chlorinated tap water decomposes it almost immediately The amount of LSD necessary to provide the average hallucinogenic experience is 100 micrograms (one tenth of a milligram). Its effects begin approximately one half-hour after oral ingestion, or virtually instantaneously after spinal injection. Psilocybin is from 100 to 150 times less potent by weight, requiring doses of 4 to 8 milligrams for the average subject. Mescaline is the least potent compound, the usual dose being 400 milligrams (some 4,000 times larger than LSD) and requiring as long as two hours to take full effect. DMT and T-9 are destroyed in the stomach—hence must be given intramuscularly, producing results within three to fifteen minutes, which abate entirely within the hour.

The first effects are a sense of subjective mystery and certain vegetative changes, which include elevation of temperature, increased blood pressure, dilation of the pupils, and more rapid respiration. Mescaline, especially, tends to produce initial nausea, which passes away entirely after the desired results are obtained. Thus one experiences his "hangover" beforehand, rather than afterward, which makes the hallucinogenic experience rather unique among drug reactions. Hyper-excitability of the nervous system follows, with transient tremors, tenseness, and occasional paresthesias (sensations on the surface of the skin). This central nervous stimulation is noticeably stronger than that produced by the amphetamines and culminates in an intense euphoria, which lasts throughout the main course of the experiment. The thoracolumbar regions are especially excited, making the genital area and the calves of the legs hypersensitive. After some time, the skin becomes damp; the mouth and throat, dry; and the muscles, slightly cramped from the prolonged stimulation.

The initial phase of the experience is almost exclusively pleasant. In the case of individuals who have anxieties or feelings of insecurity, potential weaknesses may be dramatized in such a way as to precipitate acute psychotic reactions. Intravenous injections of Thorazine (50 milligrams) or Frenquel (160 milligrams) bring almost immediate release from the depressing symptoms; residual symptoms have been known, however, to persist in rare cases for a week or more. The ordinary individual has little to fear from the hallucinogenic experience, and investigators have recorded subjects who have taken LSD several dozen times without side effects; Indians who employ peyote in their religious observations have used the drug for many years without physical or mental deterioration.

The effects of LSD, mescaline, and psilocybin may last for eight to ten hours. The spectacular phase of the experiment is over in five or six hours; the only aftereffects are muscular and nervous tension, which can be controlled with nicotinlc acid, the phenothiazine tranquilers, or barbiturate sedation. For a week or so afterward, perhaps due to the depletion of serotonin in the body, a greater than usual mental alertness is noted.

2) Wordsworth's famous "Ode: Intimations of Immortality" is perhaps the best-known expression of this suggestion of a former paradise, which adult perception fails to sustain:

> Our birth is but a sleep and a forgetting:
> The Soul that rises with us, our life's Star,
> Hath had elsewhere its setting,
> And cometh from afar:
> Not is entire forgetfulness,
> And not in utter nakedness,
> But trailing clouds of glory do we come
> From God, who is our home:
> Heaven lies about us in our infancy!
> Shades of the prison-house begin to close
> Upon the growing Boy
> But He beholds the light, and whence it flows,
> He sees it in his Joy;
> The Youth, who daily farther from the east
> Must travel, still is Nature's Priest.
> And by the vision splendid

Is on his way attended;
At length the Man perceives it die away,
And fade into the light of common day.

3)   The haunting suggestion of a former mode of existence is one of the common themes of literature in all cultures. An example is given from Gerard de Nerval's "Odelettes" (*Fantaisie*):

There is an air for which I would give all
Rossini, all of Weber and Mozart,
A very old air, languid, funereal,
Which charms me only with its secret art.
But every time I happen to hear it sung,
My soul grows younger by two centuries;
It is under Louis Treize . . . I believe I see
A green hill yellowed by the setting sun.
Then an old brick castle with stone corners,
All the windows stained with rosy colors,
Surrounded by great parks, with a little river
Bathing their feet as it glides among the flowers.
Then a lady, at the tall window of her chamber,
A blonde with dark eyes, in an ancient gown . . .
Whom I have seen before perhaps and known
In a former existence! . . . which I can remember.

4)   N.A. Yawger (*American Journal of Medical Science*, 195, 351-357, 1938) makes the following remark, regarding the ability of cannabis (also a hallucinogenic drug) to summon memories long buried in the subconscious mind:

John Stuart Mill, philosopher, wrote of its power to revive forgotten memories, and in my enquiries, smokers have frequently informed me that while under its influence, they are able to recall things long forgotten. If through such use the unconscious mind could be rendered more accessible, possibilities as an aid in psychoanalysis and psychotherapy are shown.

Drs. Rolls and Stafford-Clark (quoted by de Ropp, *Drugs and the Mind*) add this comment:

Characteristic effects of *Cannabis Indica* include euphoria with disturbances of time and space perception, sometimes accompanied by erotic visual imagery and usually followed by profound relaxation and sleep. A striking feature may be the vivid recall or re-experience of feelings long since past and formerly forgotten. This drug thus seemed particularly suited to our requirements in the treatment of (a) case of depersonalization, and proved in fact empirically to be highly successful.

5)  The need to express our own inner needs and motives is perhaps the most important aspect of fulfillment in life. Hermann Hesse's definition of the problem, which leads to neurotic loss of powers if unsolved, is as follows:

> Ich wollte ja nichts als das zu leben versuchen, was von
> selber aus mir heraus wollte. Warum war das so schwer?
> (I wanted nothing more than to
> give expression to that which I was born with.
> Why was it so hard?) (*Demian*)

Goethe's Faust also voices a similar complaint:

> Der Gott, der mir in Busen wohnt,
> Kann tief mein Innerstes erregen;
> Der uber allen meinen Kraften thront,
> Er kann nach aussen nichts bewepen.
> Und so ist mir das Dasein eine Last,
> Der Tod erwunscht, das Leben mir verhasst.
> (The God who dwells within my breast can
> deeply stir my soul; yet He who governs
> all my powers can not express himself in
> outer deeds. Hence, my existence is a burden
> to me, death would be welcome; life is cursed.)

That happiness depends upon the realization of our inmost nature is explained by the fact that the basis for reality is within us, rather than outside, and must be given expression in order keep the flow of life in motion. Consider Hesse's *Demian*:

> My young friend, you also hare your mysteries.
> I know you must have dreams which you have not

told me. I do not wish to know what they are,
but I must tell you: live these dreams, put them
into action, learn to respect them . . . We must
renew them every day within our hearts, otherwise
the world will become a sorry place.
The things which we see are the same things
which are inside of us. There is no reality
other than that reality which we bear within.
For that reason, most people live such unreal
lives, because they think the world outside is
the real world, and don't allow the world
inside of them to express itself.

6)  Also compare Fitz Hugh Ludlow, *The Hasheesh Eater*:

> In the midst of my complicated hallucination, I could
> perceive that I had a dual existence. One part of me was
> whisked unresistingly along the track of this tremendous
> experience, the other sat looking down from a height
> upon its double, observing, reasoning and serenely
> weighing all the phenomena.

Havelock Ellis also noted the intellectual clarity that is preserved
during the hallucinogenic experience:

> The mescal drinker remains calm and collected amid the
> sensory turmoil around him; his judgment is as clear as
> in the normal state; he falls into no oriental condition of
> vague and voluptuous reverie. The reason why mescal is
> of all this class of drugs the most purely intellectual in
> its appeal is evidently because if affects mainly the most
> intellectual of the senses. On this ground it is not probable
> that its use will easily develop into a habit.

7)  The ability possessed by hallucinogenic agents to transform
the commonplace into the sublime is described by Ellis in another
passage from, his study on mescaline:

> A large part of its charm lies in the halo of beauty which it
> casts around the simplest and commonest things.

The effect upon a common room was one of extreme, refreshing novelty: The difference between the room as I saw it then and the appearance it usually presents to me was the difference one may often observe between the picture of a room and the actual room. The shadows I saw were the shadows which the artist puts in, but which are not visible in the actual scene under normal conditions of casual inspection. I was reminded of the paintings of Claude Monet, and as I gazed at the scene it occurred to me that mescal perhaps produces exactly the same conditions of visual hyperaesthesia, or rather exhaustion, as may fee produced on the artist by the influence of prolonged visual attention. I wished to ascertain how the subdued and steady electric light would influence vision, and passed into the next room; but here the shadows were little marked, although the walls and floor seemed tremulous and insubstantial, and the texture of everything was heightened and enriched.

8) An unnamed artist friend of Havelock Ellis observed the same archetypal spectacle mentioned here, though he did not consciously recognize the animals as sperm cells: "At another time ray eye seemed to be turning into a vast drop of dirty water in which millions of minute creatures resembling tadpoles were in motion."

In *Lysergic Acid Dlethylamide and Mescaline in Experimental Psychology*, edited by Louis Cholden, R.S. Sandison repeats the experience of a female patient who underwent a birth reenactment:

> I have found the vaginal passage and I have the great lips of the vagina sealed . . . I must get out. I am pressing against them with my feet and hands and the seals are giving way . . . I can feel my young body starting to make another effort . . . I feel I must grow away from the womb, I feel I have left it, but not quite. I have been returning to the womb and seeing myself as a sperm swimming about, and others dying, clinging to the wall of the womb and then falling away.

9) The gaining of self-knowledge while under hallucinogenic stimulation is specifically mentioned by the Indians of the Southwest, who feel that guilt is resolved by means of unrelenting self-criticism and visionary awareness of one's sins. Ellis also refers to a similar state

of perceptive introversion: "It was as if I had unexpectedly attained an objective knowledge of my own personality. I saw, as it were, my normal state of being with the eyes of a person who sees the street on coming out of the theatre in broad day."

Harold A. Abramson (*Lysergic Acid Diethylamide and Mescaline in Experimental Psychology*) believes that this awareness of one's structural weaknesses can be beneficial by making it possible to act out the corresponding conflicts and strengthens the personality:

> Under the influence of LSD an experimental stress situation is produced. Psychodynamically, the subjects of the experiment repeatedly go through threatening situations in which they are constantly assured by their own success in dealing with the experimental stress. Ego-depression produced by the drug is well balanced by ego-enhancement which persists as an ego lesson learned.

10) Brotteaux (*Hachich; herbe de folie et de reve*) states that the final stage of hallucinogenic intoxication is "a period of ecstasy and profound tranquility". Nearly every writer on the subject agrees that the experience ends with a heightening of the sublime and the transcendent, which the subject attributes to the objects of his vision. Beringer (*Der Meskalinrausch*) says: "Extraordinary joy overcame me—a strong and beautiful feeling of eternity and infinity. This so overwhelmed me that soon everything appeared infinite."

Baudelaire (*Les paradis artificiels*) describes the culmination of the cannabis experience as follows:

> A mood of calm, muted and tranquil, takes place; the universality of man is announced colorfully, and lighted as it were by a sulfurousdawn. If perchance a vague memory reaches the soul of this poor happy man that possibly there is another God, be certain that he will rise up and question His commands and that he will face him without terror. Who is the French philosopher who said, with the intentloa of mocking modern German doctrines, "I am a god who has dined poorly"? This irony would not touch a man intoxicated by hashish. He would quietly reply: "Perhaps I did dine poorly, yet I am a god.

Ludlow, the American experimenter, records:

My mind grew solemn with the consciousness of a quickened perception. And what a solemnity is that which the hashish eater feels at such a moment. The very beating of his heart is silenced. He stands with his finger on his lip; his eye is fixed and he becomes a very statue of awful veneration. I looked abroad on fields and waters and sky, and read in them a most startling meaning. They were now grand symbols of the sublimest spiritual truths, truths never before even feebly grasped, and utterly unsuspected. Like a map, the arcana of the universe lay bare before me. I saw how every created thing not only typifies, but springs forth from some mighty spiritual law as its offspring.

11) The manner in which the outer world and the inner soul are made to correspond mystically through perception is described by Baudelaire:

Suppose you look at a tree gracefully waving in the wind; in a few seconds what, in the mind of a poet, might be merely a natural comparison, becomes for you a reality. First, you attribute to the tree your passion, your desire or your melancholy, its murmurs and its writhing become yours, and before long you are the tree. In the same way, a soaring bird first represents the immortal desire to fly above things human, but already you are yourself the bird.

This psychological phenomenon of projection is familiar to all who have investigated the matter.

# A PROVOCATIVE AND ENLIGHTING LOOK INTO OUR CONSCIOUSNESS

## The Far Off Land An attempt at a philosophical evaluation of the hallucinogenic drug experience

It has seemed to me that the well-established properties of the hallucinogenic drugs might be well employed to enable us to explore this far-off land, which is in effect our subconscious mind.
The Far-Off Land

Jo aims to bring together the perspectives of philosophy Hopefully with the immence background of anthropology, literature, comparative religion, the arts and psychology can someday be brought together with psychotropic knowledge to better understand our consciousness to ultimately improve humanity cure mental illness and even solve lifes mysteries.

Were we to learn its secrets, we would better understand our own desires, and the motives which drive us through life.

Still better, the secrets of human history would perhaps be discovered as the eternal patterns of imagination which have shaped our spiritual existence.

But perhaps most important of all, to penetrate the well of the past might restore to us that visionary perception which we think to have once possessed.

The far-off land has tremendous meaning and insight. Intelligently written and poetic. Takes you on journey that feels you full with meaning and insight that leaves you with a sense of awe and mystery attaching to our contemplation of life

Filled with historical facts and quotes from the worlds greatest minds and literature.

Eugene Seaich was a brilliant man who had 5 degrees includng a PH.D in Musicology, a Ph.D in German literature and a PH.D Philosophy and 2 Masters in Pharmacology and the Fine arts.

Has performed lectures on LSD and Psychedelics he also has writings in the University of Utah's pharmaceutical journal. And has written several books, composed 2 symphonies and is a world renowned religious scholar

www.ingramcontent.com/pod-product-compliance
Lightning Source LLC
Chambersburg PA
CBHW022118170526
45157CB00004B/1683